Molecular Recognition in Pharmacology

This unique volume traces the behavior of the drug substance, starting from the initial pre-contact stage, and ending with the formation of the complex. Molecular recognition lies in the foundation of every life form and includes many mysteries. Currently, studies on this topic in pharmacology are limited to determining the properties of complexes of medicinal substances (drugs) with specific (complementary) biomolecules: receptors, enzymes, and ion channels. This direction is very fruitful, although the phenomenon of molecular recognition is far wider. The results present the mechanisms that prevent drugs from participating in nonspecific binding.

Features

- Presents the basics of thermodynamics and kinetics of complex formation between ligands and receptors.
- Tests and validates selected novel therapeutic concepts.
- Provides a review of the pharmacophore approach and drug design methods.
- By its nature, pharmacology is a multidisciplinary science; hence, disciplinary areas include chemistry, biology, and neuroscience.

Discusses hot topics including 3D structure determination techniques and in silico methods and neural networks. The main theme of the book is the consideration of mechanisms created by nature to protect physiologically active substances from being stuck on nonspecific acceptors in the body. The book describes the materials that aid in the development of new medicinal substances. It is intended for researchers, as well as upper level undergraduate and graduate students interested in the problems of molecular pharmacology and drug design.

DRUGS AND THE PHARMACEUTICAL SCIENCES
A Series of Textbooks and Monographs

Series Editor

Anthony J. Hickey
RTI International, Research Triangle Park, USA

The Drugs and the Pharmaceutical Sciences series is designed to enable pharmaceutical scientists to stay abreast of the changing trends, advances, and innovations associated with therapeutic drugs and the area of expertise and interest that have come to be known as the pharmaceutical sciences. The body of knowledge that those working in the pharmaceutical environment have to work with, and master, has been, and continues, to expand at a rapid pace as new scientific approaches, technologies, instrumentations, clinical advances, economic factors and social needs emerge and influence the discovery, development, manufacture, commercialization, and clinical use of new agents and devices.

Recent Titles in Series

High Throughput Screening in Drug Discovery, authored by Amancio Carnero

Generic Drug Product Development: International Regulatory Requirements for Bioequivalence, Second Edition, authored by Isadore Kanfer and Leon Shargel

Aqueous Polymeric Coatings for Pharmaceutical Dosage Forms, Fourth Edition, authored by Linda A. Felton

Good Design Practices for GMP Pharmaceutical Facilities, Second Edition, authored by Terry Jacobs and Andrew A. Signore

Handbook of Bioequivalence Testing, Second Edition, authored by Sarfaraz K. Niazi

FDA Good Laboratory Practice Requirements, First Edition, authored by Graham Bunn

Continuous Pharmaceutical Processing and Process Analytical Technology, authored by Ajit Narang and Atul Dubey

Project Management for Drug Developers, authored by Joseph P. Stalder

Emerging Drug Delivery and Biomedical Engineering Technologies: Transforming Therapy, authored by Dimitrios Lamprou

RNA-seq in Drug Discovery and Development, authored by Feng Cheng and Robert Morris

Patient Safety in Developing Countries: Education, Research, Case Studies, authored by Yaser Al-Worafi

Industrial Hygiene in the Pharmaceutical and Consumer Healthcare Industries, authored by Casey Cosner

Cancer Targeting Therapies: Conventional and Advanced Perspectives, authored by Muhammad Yasir Ali, Shazia Bukhari

Molecular Recognition in Pharmacology, authored by Mikhail Darkhovskiy

For more information about this series, please visit: www.crcpress.com/Drugs-and-the-Pharmaceutical-Sciences/book-series/IHCDRUPHASCI

Molecular Recognition in Pharmacology

Mikhail Darkhovskiy

CRC Press
Taylor & Francis Group
Boca Raton London New York

CRC Press is an imprint of the
Taylor & Francis Group, an **informa** business

First edition published 2024
by CRC Press
2385 Executive Center Drive, Suite 320, Boca Raton, FL 33431

and by CRC Press
4 Park Square, Milton Park, Abingdon, Oxon, OX14 4RN

CRC Press is an imprint of Taylor & Francis Group, LLC

Library of Congress Cataloging-in-Publication Data
Names: Darkhovskiy, Mikhail, author.
Title: Molecular recognition in pharmacology / authored by Mikhail Darkhovskiy,
InterX Inc, Berkeley, CA, USA.
Description: First edition. | Boca Raton : CRC Press, 2024. | Series: Drugs and the pharmaceutical sciences |
Includes bibliographical references and index. |
Identifiers: LCCN 2023025912 (print) | LCCN 2023025913 (ebook) | ISBN 9781032431086 (hardback) |
ISBN 9781032432946 (paperback) | ISBN 9781003366669 (ebook)
Subjects: LCSH: Molecular pharmacology. | Drugs–Reactivity. | Molecular recognition.
Classification: LCC RM301.65 .D37 2024 (print) | LCC RM301.65 (ebook) |
DDC 615.7–dc23/eng/20230822
LC record available at https://lccn.loc.gov/2023025912
LC ebook record available at https://lccn.loc.gov/2023025913

ISBN: 978-1-032-43108-6 (HB)
ISBN: 978-1-032-43294-6 (PB)
ISBN: 978-1-003-36666-9 (EB)

DOI: 10.1201/9781003366669

Typeset in Times
by Newgen Publishing UK

Contents

Preface

Almost 20 years have passed since the publication of the book *Molecular Recognition: Pharmacological Aspects*, which we wrote with the respected co-authors E.N. Gorbatova and V.N. Kurochkin. The book achieved success and was being re-published. Around 2007 we realized that it is worth writing a new book to give a deeper look at many aspects of molecular recognition. The book was finished just before the COVID-19 pandemic and successful designing of a vaccine within a year, which is also based on the molecular recognition mechanism of the spike protein of the viruses by immune cells. Despite such a long time and progress in biosciences including high-throughput screening methods, fast and reliable determination of the protein–ligand crystal structure using cryo-EM, routinely used computer simulations, and machine learning boost, research on the study of molecular recognition mechanisms has remained almost the same: the main direction of research is to determine the properties of a substance complex with its specific receptor. This direction is highly productive, but the phenomenon of molecular recognition is much broader. Although there are numerous studies on molecular recognition, the mechanisms of its amazing efficiency and reliability remain insufficiently studied.

As for the complementarity principle, there is no mystery in molecular recognition. However, seeing the recognition from the other side, as a problem of how a drug avoids the "trap" of numerous nonspecific acceptors in the body, then the problem of recognition is not as simple as seemed before. It is true that the reason for avoiding nonspecific sites is based on the same complementarity principle, but this explanation does not convey the complexity of the recognition process. The fact is that a drug substance may form a stable complex with noncomplementary acceptors, for example, with graphitized carbon black, which is certainly not complementary to those. Nevertheless, the complex with graphitized carbon black may be as tight as a complex with the specific receptor. When a drug molecule is immersed in the cavity of a protein or a cell membrane, noncomplementary complex can be strong as a result of universal dispersion forces, which would cause a drug's nonspecific binding. It is important that molecular recognition is based on weak dispersion forces because it enables the fast correction of recognition errors. Pauling (Nature, 1974, v.248, p.769) came to the conclusion that regardless of the method used to describe the specificity of weak interactions between biomolecules, van der Waals interactions play a key role in them.

Drugs reach their "own" specific receptor, even when applied at considerably low doses.[1] The book focuses on factors that prevent drugs from participating in such nonspecific binding. Among them, molecular electrostatic potential, bound water molecules, and kinetics of conformational changes are singled out. Assume that the side effects of drugs are attributed to their common pharmacophoric element with the nonspecific receptors. In this regard, the "recognition signature" concept is introduced, and its difference from well-known pharmacophore is explained.

The book includes both a review of the current research status in the field and tries to formulate the common concepts and problems of molecular recognition in pharmacology. We describe the basics of thermodynamics and kinetics of drug–receptor binding and point to actual problems and advances in the fields. All chapters are written to provide a wide outlook on the known mechanisms of recognition.

For the fundamental purpose, we also address the reader to the old, yet comprehensive, monography *Enzymes and Inhibitors of Metabolism* authored by J.L. Webb, where sufficient accuracy of methods for calculating the binding energy is combined with the simplicity of the mathematical apparatus, and to classics *From Neuron to Brain*, authored by J.G. Nicholls, A.R. Martin, B.G. Wallace, and P.A. Fuchs, which may be called the encyclopedia of neurobiology.

The role of bound water molecules in recognition and the mechanisms of agonists' and antagonists' action in nicotinic, muscarinic, and opiate receptors are discussed. Examples of drugs with high selectivity and nonselective drugs, the pharmacological effect of which is attributed to a wide spectrum of affinity for various receptors, are given. As an example of nonselective drugs, neuroleptics and antidepressants are considered. The effect of conformational entropy during drug–receptor complex formation is described.

In the book, relatively older experiments are cited for binding K_d values. Nowadays, measurements of K_d are mainly performed on pure cloned receptors, and K_d values sometimes differ from the results of radioligand experiments on tissue homogenates, where the receptors are not separated from the cell membranes. Nevertheless, despite the measurements on homogenates being more consistent with those under *in vivo* conditions, experiments with cloned receptors offer a possibility to measure the affinity of drugs to receptors of various human tissues including human brain.

Yet, a lot of problems, discoveries, and tools related to molecular recognition in pharmacology remain, which is out of scope of the book. Among them, inverse agonism and constitutively active receptors, protein–protein recognition, allosteric interactions, and artificial and orphan receptors are not discussed. Nevertheless, the themes discussed in the book have been remaining at the center of molecular pharmacology research for decades. The mentioned topics are quite interesting, although of secondary importance from the main theme of the book.

Most of the considered examples and cases are taken from ligands of neuroreceptors. Apart from them, many drugs with sophisticated molecular patterns targeting other receptors and enzymes are being developed. However, we believe that the principles of recognition and formation of complexes with receptors are common for the design of drugs of any structural type.

In writing the book, we followed the well-known principle, "no numerical method can reveal what was not measured in the experiment before," so, we tried to substantiate computational and computer models with experimental data.

Opinions presented in the book might be debatable, and the author would gratefully accept comments and criticism. Any commentaries and reflections on the topics lying within the book scope and related to molecular recognition in general are welcome.

We devote the book to the memory of our colleagues Elena Gorbatova and Vladimir Kurochkin.

NOTE

1   We do not consider metabolic breakdown of drugs in this regard.

Acknowledgments

During the COVID-19 pandemic, when the book had been already finished, its founding author, F.S. Dukhovich, passed away. I owe much to him, as he introduced me, a pure quantum chemist and molecular modeler, into a world of molecular pharmacology and taught me to see molecular recognition as rooted in biological evolution. The book was our mutual business. Now, it is a tribute to the memory of Prof. Dukhovich.

I express my deep gratitude to colleagues who helped me in writing this book. I am thankful to Prof. Sergey Varfolomeev and Prof. Andrey Rubin for proofreading this manuscript and providing valuable comments.

I am grateful to colleagues from InterX for the fruitful discussions about drug design principles and practice of detailed consideration of molecular mechanisms. I am thankful to my wife Yulia for constant support, and to my daughter Maria for inspiring me to be creative and audacious.

Finally, the generous support of the book Editor Hilary Lafoe, as well as Editorial Assistant Sukirti Singh is kindly acknowledged.

Author

Mikhail Borisovich Darkhovskiy graduated from the Moscow State University in 1996 with a major in Quantum Chemistry. In 2005 he earned a PhD in Chemistry and Solid-State Physics under the supervision of Dr. Andrei Thougreeff. His research was in electronic structure and transformation of spin-isomers of iron and cobalt complexes in spin transitions using the QM/MM effective crystal field method. Together with Prof. Dukhovich, he investigated the factors involved in molecular recognition in "ligand–receptor" systems, the influence of water on complex strength, and conformational entropy as a hidden factor for complex formation. Currently, he works in the biotech startup InterX (a subsidiary of NeoTX company), participating in the development of methods for the accurate prediction of compounds' binding affinity to proteins. His awards include the INTAS Young Scientist award, and, Emory University Visiting Scientist award (Cherry L. Emerson Center for Scientific Computation Fellowship).

Acronyms and Abbreviations

ACh	acetylcholine
AChE	acetylcholinesterase
BuChE	butyrylcholinesterase
D	Debye (measure)
D_1, D_2	subtypes of dopamine receptors
d	intermolecular distance
d_e	equilibrium intermolecular distance defined by VDW radii of contact atoms
E	potential energy of interacting molecules or groups
ΔG	change of the Gibbs free energy
H_1	subtype of histamine receptors
$5\text{-}HT_1$, $5\text{-}HT_2$	subtypes of serotonin receptors
$[I_{50}]$	concentration producing 50% of the inhibition effect
K_d, Kd	dissociation constant
K_e	equilibrium constant
K_a	affinity constant
k_B	Boltzmann constant
k_{+1}, k_{-1}	rate constants of a complex formation and dissociation, respectively
L	ligand
LR	ligand–receptor complex
mAChR	muscarinic acetylcholine receptor
nAChR	nicotinic acetylcholine receptor
q	charge
QNB	3-quinuclidinyl benzilate
R	receptor
R	universal gas constant
R	molar refraction
TMA	tetramethylammonium cation
t	average lifetime of a complex (ligand residence time)
$t_{1/2}$	half-lifetime of a complex
vdW, VDW	van der Waals
W	Water molecule
Z	number of electrons on valence orbitals
Z	number of molecular collisions per second
z	ion valence
α	molecular polarizability
α_1, α_2	subtypes of adrenaline receptors
μ	dipole moment
χ	Debye-Hückel constant

Introduction

Molecular recognition and new drug development: Extracts from conducted studies

The book begins with a section which is usually given as a conclusion. This is done so that readers can easily determine how much the book matter is fitted to their interests. Rules and propositions are briefly presented regarding the development of novel medicinal drugs selectively acting on neuroreceptors. These rules are derived from the following paradigm: 1) The central direction in investigation of molecular recognition would be to clarify mechanisms protecting drugs from non-specific binding in an organism. 2) Formation of a drug complex with the specific receptor that is based on the principle of complementarity is the result of recognition, its final step.

Let us outline some heuristics and rules for constructing drug molecules that are valid for compounds with a molecular weight of no more than 800 Da. It includes at least half of all medicinal compounds produced thus far:

1. Protection of the drug from nonspecific binding in an organism is realized through the effect of molecular electrostatic potential (MEP), bound water molecules, and rate of conformational changes of both the drug molecule and the receptor.
2. The MEP-type receptor is defined as a combination of the number and sign of the charges of ionic groups (might be zero when there is no charged group) in the orthosteric binding site (i.e., receptor's active center). In addition to the specific receptor, medicinal compounds may display affinity to receptors of the same MEP type as the target receptor.
3. An increase in the drug affinity toward the target receptor does not always result in the selectivity increase.
4. The limit affinity of a drug to the specific receptor is represented by $K_d = 0.01$ nM. Compounds with $K_d \leq 0.01$ nM likely belong to the category of toxins.
5. There are no receptors with affinity to a specific ligand $K_d \leq 0.01$ pM.
6. The number of charged groups in the drug molecule and in the binding pocket of the specific receptor must be the same to achieve proper binding. If there are no charged groups in the drug with the therapeutic effect, then the binding pocket of the specific receptor will have no charged groups.
7. The charge of a single ionic group in the drug molecule and in the binding site of the receptor cannot be more than one electron charge ("+1" or "−1").
8. The activity of the drug can be controlled/modified while preserving its selectivity by modifying the groups that are not in the composition of the recognition signature (the concept is discussed in Chapter 9).
9. The search for drugs with a new mode of action might be carried out by the determination of missed recognition signatures.
10. A linear regression equation between the dissociation rate constant (k_{-1}) and the K_d value can be used for the estimation of ligand residence time for receptor ligands, enzyme substrates and inhibitors, and ion channel blockers with a molecular weight of less than 600 Da.

DOI: 10.1201/9781003366669-1

11. The dissociation rate constant k_{-1} (or residence time) of a ligand is determined by the number of contact points in the complex with the receptor. The higher the number of contacts, the smaller is the k_{-1} value and the larger is the residence time.

12. Nitrogen atoms in amino alkyl groups of medicinal compounds that are specific to a certain type of receptors with single-charged binding site should be separated by no more than three C-C bonds.

13. The agonist paradox—that is, the agonist's ability to cause significant and pronounced conformational changes in the receptor when compared with small changes due to partial agonists' and antagonists' action, while the measured binding free energy is similar in both cases—is explained by the fact that the high-energy steps of interaction of the agonist with the receptor are "hidden" because of its fast dynamics.

14. Fast dissociation of the ACh–nAChR complex occurs not only due to the decrease in the mediator's concentration in synapse because of the action of acetylcholinesterase but also due to the increase in the local dielectric constant when water permeates into the area of ion–ion interaction, due to conformational changes in the complex.

15. Continual strengthening of some quinuclidine glycolates leads to binding with mAChR, while the free compound is almost completely absent in systemic circulation; this aspect can be explained by the formation of carbocation and the release of the water molecule from the receptor site. Strengthening of the complex with the receptor occurs because the interaction in the formed ion pair comes closer together. However, the effect of allosteric sites in the two-stage kinetics cannot be excluded.

16. Antidepressants should have low affinity to D_2-dopamine receptors.

17. Conformational entropy is a fundamental factor protecting receptors from irreversible inactivation during complex formation with compounds that have a relatively large molecular size. Enkephalins, endorphins, hormones, peptide-like therapeutics, and others belong to this type of compounds.

18. The new mechanism of action of partial agonists on mAChR receptors is proposed.

1 Energy of Interactions in Drug–Receptor Complex Formation

1.1 PROPERTIES OF COMPLEXES

Chemical substances can be used as drugs if they exert selective effects on biological targets such as receptors, enzymes, nucleic acids, ion channels, and others. Generally, the selectivity is conditional upon the properties of drug–receptor complexes.

A compounded structure comprising a group of molecules can be named as a complex if the molecules preserve their original structures even after the dissociation of the complex. The basic properties of complexes are outlined as follows:

- The bonds that participate in complex formation do not involve electron transfer between atoms, which is in contrast to covalent (chemical) bonds.
- Complexes between drugs and their biological targets are formed through multiple weak intermolecular interactions, with the total energy not exceeding 15 kcal/mol. The energy of a single covalent bond is in the range of 50–100 kcal/mol.
- The intermolecular distances in complexes can vary greatly as compared to covalent bonds with certain interatomic distances and bond angles.
- In a molecule, the number of bonds is constant, whereas in a complex, it is quite difficult to evaluate the number of intermolecular contacts (bonds). As for complexes, the term "bond" has a conditional meaning: it is better to use "van der Waals bonds" or vdW bonds, "intermolecular interactions," and "binding."
- A ligand can be displaced from the complex by applying an agent with competitive action. By contrast, it is impossible to break even a weak covalent bond through this approach.
- Stoichiometry is an inappropriate term when referring to interactions in complex formation, comparing its importance to chemical reactions.

All molecules in an organism exist as complexes with biomolecules, water, or ion sphere components. In the absence of specific ligands, the active sites of enzymes or receptors are bound to foreign substances that are unable to produce a pharmacological response or induce an enzymatic reaction.

Complex formation precedes biochemical reactions and governs the specificity of the complex. For example, complex formation precedes covalent bonding when the antibiotics exert their action at the target site, specifically in cells of pathogenic microorganisms.

It is commonly accepted that selectivity is based on complementarity mechanisms, wherein electronic and steric matching of molecules or their parts (active sites) is involved in complex formation. As complementary complexes are more stable, sometimes there is a misconception that complementary molecules participate in binding but noncomplementary molecules do not. Noncomplementary molecules also bind. A typical example of unspecific and stable complex formation is the adsorption

DOI: 10.1201/9781003366669-2

of substances on carbon graphite (see further discussion in Chapter 5). However, mechanisms have been developed to reduce the formation of stable complexes between mismatching partners in organisms.

The amount of energy required for complex formation can be evaluated using the potential energy of electrostatic interactions. Mostly, complex formation energy (complex strength) is expressed by a change in the Gibbs free energy equal to the maximum work in an equilibrium process at constant temperature and pressure:

$$\Delta G = \Delta H - T\Delta S$$

where ΔH is the enthalpy change, T is the absolute temperature, and ΔS is the entropy change.

Potential energy characterizing heat dissipation is described in enthalpy terms in the equation. Potential and free energies are expressed in Joules (J) or calories (Cal) per mole

$$(1 \text{ Cal/mol} = 4.18 \text{ J/mol})$$

1.2   EQUATIONS USED TO CALCULATE THE ENERGY OF INTERMOLECULAR INTERACTIONS

The major sources of energy for complex formation are electrostatic or electrokinetic attractive forces. Electrostatic interactions are those that exist between ions, ions and dipoles, dipoles, ions and induced dipoles, dipoles and induced dipoles, and hydrogen bonding. Electrokinetic forces, also known as London forces, are involved in dispersion interactions that are versatile for all atoms and molecules. Universal forces that mainly influence the energy of complex formation at equilibrium distances are the repulsion forces of molecular electronic shells (exchange interaction).

Although several published studies have proposed equations to calculate the potential interaction energy, it is worth describing them again in this chapter. The book *Enzyme and metabolic inhibitors*, authored by J.L. Webb, which describes the methodology of a precise and relatively simple evaluation of ligand–protein binding energy, is used herein comprehensively.

Tables 1.1 and 1.2 presents the covalent and van der Waals (vdW) radii of atoms and groups, which are used to estimate minimal approachable (equilibrium) intermolecular distances.

TABLE 1.1
Covalent and van der Waals Radii of Atoms [1]

Atom	Covalent Radius, Å		van der Waals Radius, Å
	Single Bond	**Double Bond**	**van der Waals Radius, Å**
H	0.30	—	1.2
C	0.77	0.67	1.57
N	0.70	0.61	1.5
O	0.66	0.57	1.4
S	1.04	0.95	1.85
P	1.10	1.00	1.9
F	0.64	0.55	1.35
Cl	0.99	0.90	1.80
Br	1.14	1.05	1.95
I	1.33	1.25	2.15

TABLE 1.2
van der Waals Radii of Groups and Molecules [1]

Group	van der Waals Radius, Å	Remarks
-OH	1.40	from O, head-on
	2.16	from O, along O-H
-SH	1.85	from S, head-on
	2.54	from S, along S-H
-CH$_3$	1.57	from C, head-on and side-on
	2.27	from C, along C-H
-NH$_2$	1.50	from N, head-on
	2.20	from N, along N-H
-N$^+$H$_3$	1.70	from N, head-on
	1.98	from N, side-on
	2.20	from N, along N-H
-COO$^-$	1.40	from O, head-on
	2.15	from C, side-on
	2.36	from C, along C-O
-N$^+$(CH$_3$)$_3$	3.17	from N, head-on
	3.74	from N, along N-CH$_3$
-CH$_2$-	2.10 × 2.23	half-thickness of polymethylene chain
C$_6$H$_6$	1.85, 1.7 [2]	half-thickness
H$_2$O	1.40	from O, closest approach
	2.16	from O, along O-H

1.3 ION–ION INTERACTION

Many pharmaceutical agents bind to receptors in the ionic form. The electric force, F, acting in a given space point and induced by the charge is

$$F = \frac{q}{d^2\varepsilon} = \frac{ze}{d^2\varepsilon}, \qquad (1.1)$$

where q is the charge value, d is the distance between the charge center and the given point, z is the ion valence, e is the electron charge, ε is the dielectric constant of the environment.

The charge located in the field of another charge is acted upon by the force $\mathbf{f}$ with modulo:

$$f = qF = \frac{q_1 q_2}{d^2\varepsilon} = \frac{z_1 z_2 e^2}{d^2\varepsilon} \qquad (1.2)$$

The Coulomb interaction energy between two ions is

$$E = \frac{z_1 z_2 e^2}{d\varepsilon}. \qquad (1.3)$$

Opposite charges attract each other and the same charges repel; therefore, the attraction energy has a "−" sign and the repulsive energy has a "+" sign. Noteworthy, the binding energy (E) decreases with strengthening of complexes and vice versa. When the charge of an electron (4.8·10^{-10} e.s.u.)

is separated by a distance of 1 Å, the interaction energy of 1 M ion pairs ($6.02 \cdot 10^{23}$ pairs) is thus derived using the following equation:

$$E = 332 \frac{z_1 z_2}{d\varepsilon} \text{ kcal / mol },\tag{1.4}$$

which can be rewritten for monovalent ions as

$$E = \frac{332}{d\varepsilon} \text{ kcal / mol},\tag{1.5}$$

When ions are separated by an equilibrium distance (d_e), equal to the sum of vdW radii, the interaction energy with correction for the repulsion of electronic shells is

$$E = \frac{305}{d_e \varepsilon} \quad \text{ kcal / mol},\tag{1.6}$$

The attraction energy of two monovalent ions with opposite charges at 10 Å is −33.2 kcal/mol in vacuum, while it is less than −1 kcal/mol in an aqueous solution.

1.3.1 Cationic Groups in Drugs

Cationic groups in drug molecules do not differ widely and contain nitrogen as a common element. In primary, secondary, and tertiary amines, the nitrogen atom readily accepts a proton at pH 7.2–7.4. The N atom, while achieving a full octet of electrons in its outer electron shells, has a lone pair of electrons to form a new dative covalent bond with a proton. Upon sharing the donor's electron pair, the nitrogen atom gains a positive charge, numerically equal to the charge of one electron.

The structure of a cationic group in many drugs is depicted in Figure 1.1, where R_1 is the backbone of the molecule, and R_2 and R_3 represent H, CH_3, or C_2H_5 groups in most cases.

Cationic groups are often present in cyclic moieties such as benzoclidine (Figure 1.2). Some drugs contain cationic groups formed upon the protonation of amidine in heterocyclic or guanidine groups (Figure 1.3). Examples of drugs with these cationic groups are naphazoline and guanfacine (Figure 1.4).

As a rule, cationic groups present in drug molecules that do not affect the central nervous system are presented by a quaternary nitrogen atom (Figure 1.5).

The charge is not concentrated on the nitrogen atom but rather distributed on adjacent carbon atoms—for example, tetramethylammonium and protonated trimethylamine (Figure 1.6). However, the nitrogen atom can be considered conventionally as the geometric center of a charge group[1].

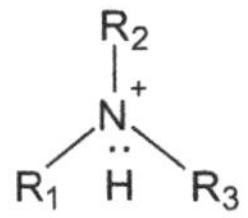

FIGURE 1.1 Structure of typical drug cationic group.

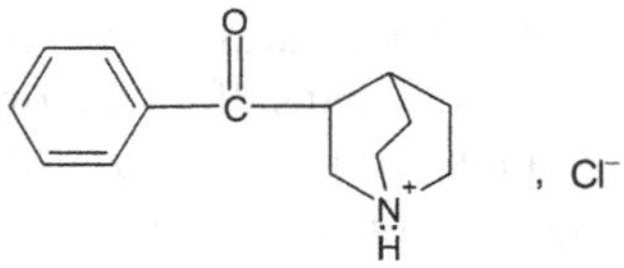

FIGURE 1.2 Benzoclidine.

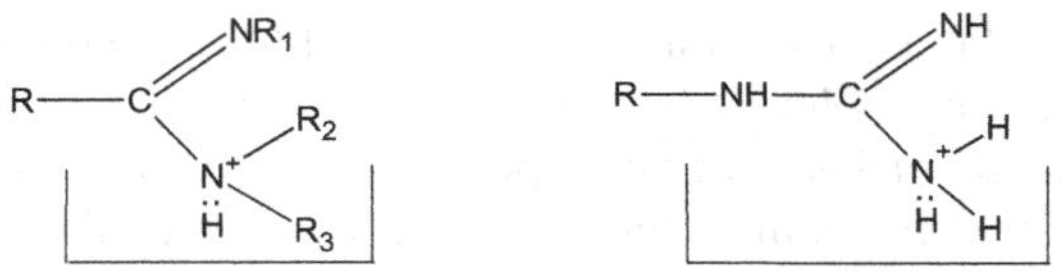

FIGURE 1.3 Typical cationic groups in drug molecules. Amidinium (1) and guanidinium (2) cationic groups are marked by frames, R is the major carbon chain, R_1, R_2, and R_3 refer to hydrogen or other substituents.

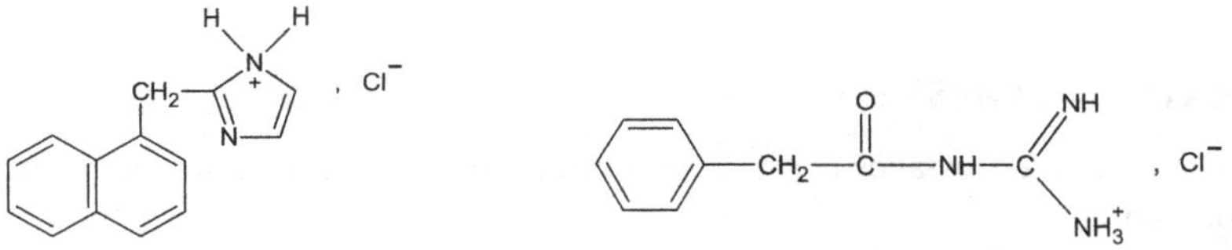

Naphazoline   Guanfacine

FIGURE 1.4 Drug molecules with amidinium and guanidinium cationic groups.

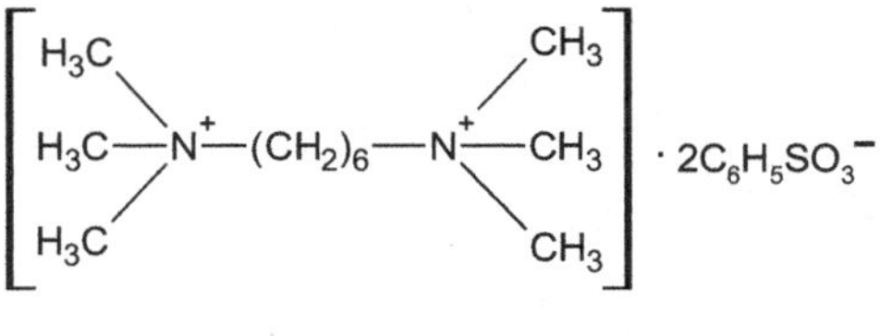

FIGURE 1.5 Hexamethonium benzosulfonate.

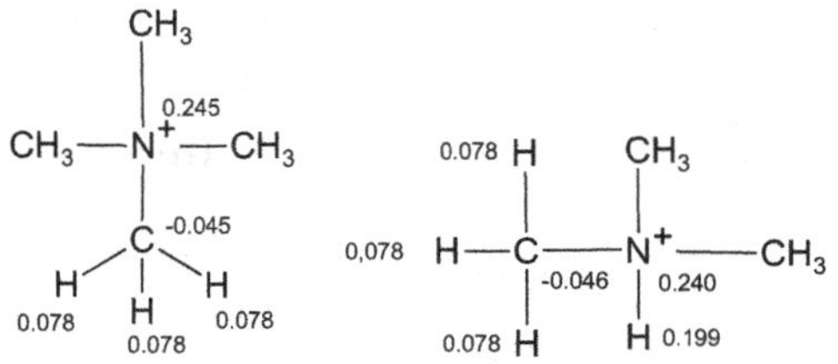

TMA cation   Protonated trimethylamine

FIGURE 1.6 Charge distribution in protonated amines.[1]

The atoms O, S, and Cl in molecules of different drugs have lone electron pairs but are unable to accept a proton in neutral media, and oxonium and sulfonium ions can be formed only in a strongly acidic medium. The higher mobility of a lone pair in the nitrogen atom than in the oxygen atom can be explained by a smaller nuclear charge and lower electronegativity.

1.3.2 ANIONIC GROUPS IN DRUGS

Typically, the dissociated carboxyl group is the main anionic group present in drug molecules. An uncompensated charge equal to the electron charge is typical for this group. Mostly, it is present as an anionic component of a drug molecule bound to the cationic center of the receptor's active site. Representative drugs or amino acids with this type of ion–ion interaction are aspirin, diclofenac, prostaglandins (Figure 1.7), gamma-aminobutyric acid, glycine, and others.

The hydroxyl group (OH–) may also function as an anionic group. Because of the mesomeric effect, electron-acceptor substituents decrease the partial charge of oxygen that favor proton cleavage. Ascorbic acid (Figure 1.10) and muscimol (Figure 1.11) without carboxylic acid form such anions.

Charge distribution in the acetate anion has two equivalent oxygen atoms as shown in Figure 1.8. The OH-, CH_2-, and CH-groups may acquire acidic properties under the effect of electron-acceptor substituents at physiological pH.

1.4 IONIZATION CONSTANTS

Basicity, or the ability of a substance to attach protons, is quantitatively expressed by the equilibrium constant K_b for a base:

$$B + H_2O \rightleftarrows BH^+ + OH^-$$

$$K_b = \frac{a_{BH^+} \times a_{OH^-}}{a_B \times a_{H_2O}}$$

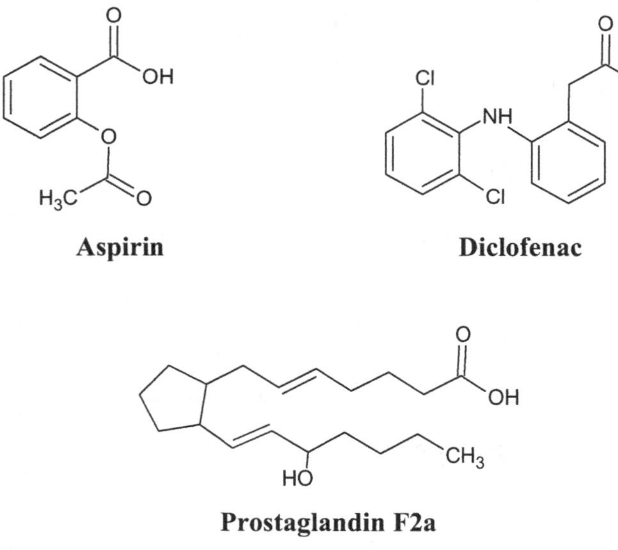

Aspirin   **Diclofenac**

Prostaglandin F2a

FIGURE 1.7   Drug molecules with an anionic group.

where B is a base and a is the thermodynamic activity. As for diluted media, activities can be equated to true concentrations.

It is convenient to estimate the strength of bases using the pH scale, which is common for both acids and bases. According to the Brønsted-Lowry theory, all proton-donating compounds, including methane and benzene, as well as protonated base BH^+, are referred to as acids. The strength of acids is quantitatively measured based on the acid dissociation constant K_a for the equilibrium state:

$$AH + H_2O \rightleftarrows A^- + H_3O^+$$

$$K_a = \frac{\left[A^-\right]\left[H_3O^+\right]}{\left[AH\right]\left[H_2O\right]}$$

where AH is an acid and A^- is a base.

Considering that the $[H_2O]$ value is constant for diluted media and numerically equal to 1, we have

$$K_a = \frac{\left[A^-\right]\left[H_3O^+\right]}{\left[AH\right]}$$

Compared with K_a, its negative logarithm, pK_a, is more useful. pK_a is the pH at which 50% of an acid is ionized. The lower the pK_a value, the stronger the acid is. Acids are thus classified as very strong ($pK_a < 0$), strong ($0 < pK_a < 4.5$), weak ($9 < pK_a < 14$), and very weak ($pK_a > 14$). Contrarily, the higher the $pK_a > 14$, the stronger the base is ($pK_a > 14$, very strong; $9 < pK_a < 14$, strong).

The degree of ionization (J) of a base in the aqueous solution is calculated as

$$J(\%) = \frac{100}{1 + \log^{-1}\left(pH - pK_a\right)}$$

Samples of the J value are presented in Table 1.3.

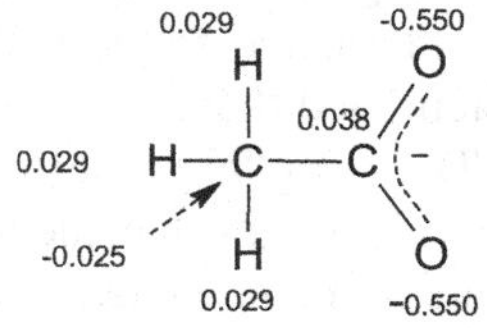

FIGURE 1.8 Charge distribution in the acetic anion.

TABLE 1.3
Degree of Ionization Based on pH and pK_a [3]

$pK_a - pH$	Ionization of Anions, %	Ionization of Cations, %
−4	99.99	0.01
−3	99.94	0.1
−2	99.01	0.99
−1	90.91	9.09
0	50.00	50.00
1	9.09	90.91
2	0.99	99.01
3	0.10	99.94
4	0.01	99.99

Drugs containing positively and negatively charged molecules cannot penetrate the blood–brain barrier (BBB) or enter a cell through the cytoplasmic membrane because ions that are surrounded by hydrate shells cannot penetrate the hydrophobic membrane barriers. When a drug is present both in the charged and neutral form in living organisms, it easily crosses the blood–tissue barriers even at very small concentrations of the neutral form, mainly attributed to the shift in the equilibrium toward the formation of an uncharged ion form during its removal. This process occurs until the equilibrium concentration is established on both sides of the barrier.

Drugs containing an ammonium group with the quaternary nitrogen atom, have a constant charge that is independent of pH; as they cannot penetrate the BBB, they display their effect only at the peripheral receptors. Because of this aspect, these drugs are referred to as myorelaxants, bronchodilators, ganglioplegic agents, and others.

1.5 EFFECT OF SUBSTITUTING GROUPS IN MOLECULES ON IONIZATION: INDUCTIVE AND MESOMERIC EFFECTS

The type of receptor to which a drug binds to and the nature of side effects caused are both dependent on whether the drug exists in the charged or neutral form. The ability of a substance to undergo ionization is greatly influenced by its substituting groups in its molecule (Table 1.4).

When dissimilar atoms are present in a molecule, electrons participating in bond formation are shifted to a more electronegative atom. The polarization direction depends on the electronegativity of the atoms, the values of which are shown in Table 1.5.

An electron shift along the σ-bonds (the inductive effect) is described with the example of chloroacetic acid. The presence of an electron-withdrawing chlorine atom increases the strength of acids: monochloroacetic acid (pK_a 2.9) is much stronger than acetic acid (pK_a 4.8). The inductive effect drastically decreases as the distance between the chlorine atom and the carboxyl group increases because σ-electrons have low mobility: thus, γ-chlorobutyric acid has a similar strength as that of butyric acid. Taft constants are generally used to estimate the substituents' inductive effect in compounds of the aliphatic series [10].

All bonds are polarized, even those with a different type of hybridization, such as in propylene: the sp^2-hybridized carbon atom (2.8) is more negative than the sp^3-hybridized carbon (2.5).

In molecules containing conjugated alternative single and multiple bonds, the π electrons are shifted along the entire conjugated system, and therefore, fractional positive and negative charges appear on both ends of the molecules. The term "mesomeric effect" refers to the polarizing effect of an atom or a group of atoms that causes the displacement of π-electrons in conjugated bonds. Electron donor groups that can contribute to a conjugated system either completely or partially by donating an electron pair have a positive (+M) effect. By contrast, electron acceptor substituent groups have a negative (−M) effect. As π-electrons are more mobile than σ-electrons, the mesomeric effect can occur in relatively extended π-systems.

The substituents $-NH_2$, $-N(Alk)_2$, $-COO^-$, $-OCH_3$, and $-CH_3$ donate electrons, thus leading to an increase in the basicity of the other conjugated groups in a molecule. The electron-withdrawing groups $-NH^+_3$, $-N^+(Alk)_3$, $-NO_2$, $-SO_3^-$, $-HC=O$, $-COOH$, halogens, and phenyl groups accept electrons, thus leading to a decrease in the basicity (i.e., increasing the acidity) of the other conjugated groups.

For the first time, the electronic effect of substituents in aromatic compounds was estimated quantitatively by Hammett [11]. The shifting effect (illustrated in Figure 1.9) for the methyl ester group of n-aminobenzoic acid, reduces electron density on the nitrogen atom, leading to a decrease in the basicity of the amino group (pK_a = 4.63) as compared with that of aniline (pK_a = 2.38). This effect can be achieved by electron-withdrawing substituents at the p-position.

TABLE 1.4
Effect of Substituents on Acidity (Basicity) [4–8]

Conjugate Acid	Conjugate Base	pK$_a$	Conjugate Acid	Conjugate Base	pK$_a$
CH$_3$COOH Acetic acid	CH$_3$COO$^-$	5	CH$_2$NH$_3^+$	CH$_2$NH$_2$ Benzylamine	9.33
ClCH$_2$COOH Monochloroacetic acid	ClCH$_2$COO$^-$	2.8	(pyridinium) $^+$N–H	Pyridine	5.21
CH$_3$CH$_2$CH$_2$COOH Butyric acid	CH$_3$CH$_2$CH$_2$COO$^-$	4.8	(pyrrole) $^+$N–H$_2$	Pyrrole	−0.28
ClCH$_2$CH$_2$CH$_2$COOH γ-Chlorobutyric acid	ClCH$_2$CH$_2$CH$_2$COO$^-$	4.5	OH Phenol	O$^-$	9.99
RNH$_3^+$ R$_2$NH$_2^+$ R$_3$NH$^+$	RNH$_2$ R$_2$NH R$_3$N Alkylated amines	10–11	OH NO$_2$ p-Nitrophenol	O$^-$ NO$_2$	7.2
H$_2$NCH$_2$CH$_2$NH$_3^+$ $^+$H$_3$NCH$_2$CH$_2$NH$_3^+$	H$_2$NCH$_2$CH$_2$NH$_2$ Ethylenediamine	10.0 (1) 7.0 (2)	NO$_2$ OH NO$_2$ 2,4-Dinitrophenol	O$^-$ NO$_2$ NO$_2$	4
H$_2$N(CH$_2$)$_8$NH$_3^+$ $^+$H$_3$N(CH$_2$)$_8$NH$_3^+$	H$_2$N(CH$_2$)$_8$NH$_2$ Octamethylenediamine	11.0 (1) 10.1 (2)	CH$_4$	CH$_3^-$	40

(continued)

TABLE 1.4 (Continued)
Effect of Substituents on Acidity (Basicity) [4–8]

Conjugate Acid	Conjugate Base	pKa	Conjugate Acid	Conjugate Base	pKa
(piperazinium)	Piperazine	0.83 (1) 5.56 (2)	CH_3NO_2 Nitromethane	$^-CH_2NO_2$	10.2
$CH_3-C(OH^+)NH_2$	$CH_3-C(O)NH_2$ Acetamide	-0.51 -1	$CH_2(NO_2)_2$ Dinitromethane	$^-CH(NO_2)_2$	3.6
$CH_3CH_2NO_2$ Nitroethane	$CH_3CH^-NO_2$	8.6	Benzoic acid (PhCOOH)	(PhCOO$^-$)	4.20
Salicylic acid (COOH, OH)	Salicylic acid (ionized) (COO$^-$, O$^-$)	13.4	p-Aminobenzoic acid (H_2N–COOH)	(H_2N–COO$^-$)	4.92
(anilinium NH_3^+)	Aniline (NH_2)	4.63	CH_3CH_2OH Ethanol	$CH_3CH_2O^-$	18
Methyl ester of p-aminobenzoic acid (H_3N^+–COOCH$_3$)	Methyl ester of p-aminobenzoic acid (H_2N–COOCH$_3$)	2.38	C_2H_5-OH$^+$-C_2H_5	C_2H_5-O-C_2H_5 Diethyl ether	-3.59

Values of pKa >15 and <2 are considered as rough approximations.

TABLE 1.5
Electronegativity of Atoms by Pauling [9] (in Relative Scale)

Atom	Electronegativity	Atom	Electronegativity
H	2.1	Cl	3.0
C	2.5	Br	2.8
N	3.0	I	2.5
O	3.5	Na	0.9
S	2.5	K	0.8
P	4.0	Ca	1.0
F	2.1		

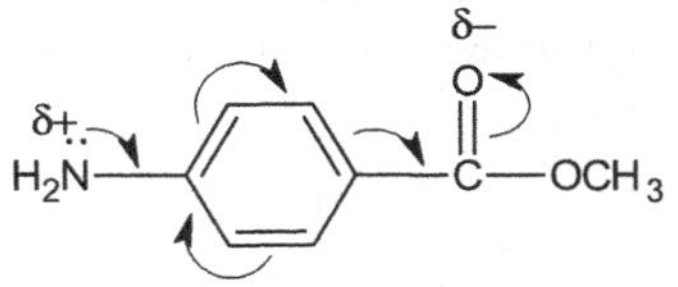

FIGURE 1.9 Mesomeric effect in the methyl ester group of n-aminobenzoic acid.

Although a lone electron pair in the nitrogen atom of pyridine is not involved in the aromatic ring, its shift causes a decline in the basicity ($pK_a = 5.21$) as compared to aliphatic amines. When the lone electron pair participates in the formation of the aromatic system, as in the case of pyrrole, the basicity further reduces ($pK_a = 0.28$). Intermediate-length CH_2 chains weaken the effect of the acceptor or donor groups. Thus, the basicity of benzylamine ($pK_a = 9.33$) is greater than that of aniline.

Methane ($pK_a = {\sim}40$) can be considered as an acid only theoretically; by contrast, nitromethane ($pK_a = 10.6$) is regarded as a weak acid and dinitromethane ($pK_a = 3.6$) as a strong acid. The (−M) effect of the carbonyl group is responsible for the reduced basicity of the amino group in amides, wherein a proton is attached mainly to an oxygen atom.

Considering the conjugation effect, the OH-group is more acidic in phenol ($pK_a = 9.99$) than in aliphatic alcohols (pK_a of ethanol = 18).

A typical example for the mesomeric effect of electron-donating groups is benzoic acid and its derivative. The amino group that increases the electron density on the O-H bond causes decreased acidity (pK_a of benzoic acid = 4.20 and pK_a of p-aminobenzoic acid = 4.92). In turn, the -COO⁻ donor group greatly reduces the acidity of the phenolic OH-group (Table 1.4).

Indeed, the mesomeric effect contributes to the acidity of vitamin C (also known as ascorbic acid; $pK_a = 4.2$). The pH of its 0.1 N aqueous solution is 2.2. The acidic properties of vitamin C are attributed to the removal of a proton from the OH-group at the C3 position in the heterocyclic moiety.

Another typical example is muscimol; it has a molecular structure distinct from that of gamma-aminobutyric acid (GABA) and exerts the activity of the GABA-receptor agonist owing to the electron-withdrawing properties of oxazole, causing deprotonation of the hydroxyl group of muscimol in the aqueous solution and its conversion into a zwitterion (Figure 1.11).

In addition to inductive and mesomeric effects, the electrostatic field influences the basicity (acidity) as with the protonation of polymethylene diamines $H_2N(CH_2)_nNH_2$. These amino groups are not the same in the context that the protonation of one amine hinders the protonation of another amine. For ethylene diamine with two CH_2 groups, the pK_{a1} and pK_{a2} values are 10.23

and 7.48, respectively. Although the field effect weakens as the distance (the number of CH_2-groups) separating the amino groups increases, it occurs even at a distance of ~11 Å between the charge centers in protonated $H_2N(CH_2)_8NH_2$ as judged from the pK values (pK_{a1} = 11.0, pK_{a2} = 10.10) [12].

The first and second ionization constants of aliphatic dicarboxylic acids $HOOC(CH_2)_nCOOH$ are different from each other, as one ionized group prevents the dissociation of another carboxyl group: the pK_{a1} and pK_{a2} values of oxalic acid are 4.19 and 1.23, respectively. Similarly, the electrostatic field effect contributes to the ionization of amino acids.

1.6   ION–DIPOLE INTERACTION

The potential energy of the ion–dipole interaction is calculated as the sum of interactions between ions and dipole charges. As the components of this sum have opposite signs, the attraction appears only at certain mutual positions of interacting charged particles. The ion–dipole interaction energy is expressed using the following equation (notations are defined in Figure 1.12):

$$E = \frac{zeq}{\varepsilon} \cdot \left(\frac{1}{d_1} - \frac{1}{d_2} \right).$$

(1.7)

Equation (1.7) gives the difference between the attraction of unlike charges and the repulsion of like charges.

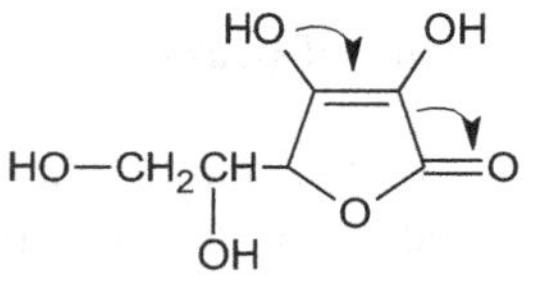

FIGURE 1.10   Mesomeric effect in ascorbic acid.

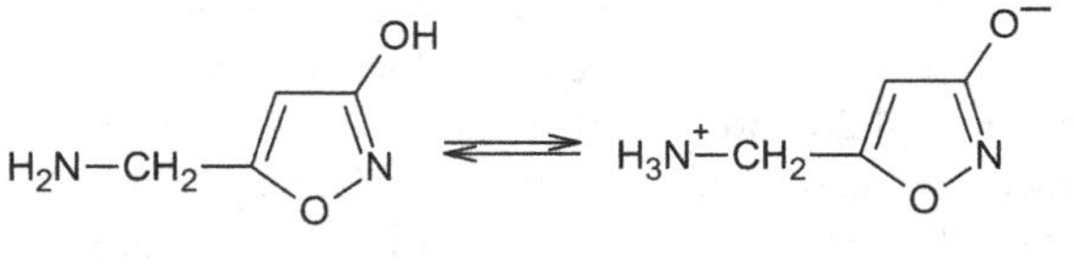

FIGURE 1.11   Muscimol zwitterion.

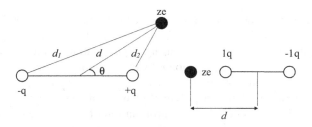

FIGURE 1.12   Scheme of ion–dipole interaction (d is the distance from an ion to the dipole's center).

If $d_1=d_2$, E=0, and the ion or dipole has the same charge as each other, then the attractive (or repulsive) force is maximal. Equation (1.7) can be transformed as

$$E = \frac{ze\mu}{d^2\varepsilon}\cos\theta \tag{1.8}$$

where $\mu = q \cdot l$, that is, μ is the dipole moment and l is the distance between the charge centers of the dipole.

If charges are located at equilibrium or very similar intermolecular distances, then it is necessary to consider the dipole moment of specific bonds or dipole groups, not the dipole moment of the whole molecule. The ion–dipole interaction may contribute substantially to the complex formation energy even if the total dipole moment of a molecule is equal to zero.

Values of dipole moments are expressed in Debye (D). Assuming that each charge of a dipole moment is equivalent to one-third of an electron charge ($1.6 \cdot 10^{-11}$ esu) and that the charge centers are present at a distance of 1.5 Å, then $\mu = 2.4 \cdot 10^{-18}$ esu. Given that 1 Debye is equal to $1 \cdot 10^{-18}$ esu, $\mu = 2.4$ D.

Equation (1.8) can be rewritten as follows:

$$E = 69.1\frac{z\mu}{d^2\varepsilon}\cos\theta \qquad \text{kcal / mol} \tag{1.9}$$

where z is the ion valence, μ is the dipole moment (expressed in Debye), and d is the distance (expressed in Å). For an equilibrium distance, d_e, Equation (1.9) is corrected for electron shell repulsion and is re-written as

$$E = 57.4\frac{z\mu}{d_e^2\varepsilon}\cos\theta \qquad \text{kcal/mol} \tag{1.10}$$

A dipole moment may occur at different positions relative to the ion. However, if the ion–dipole interaction energy exceeds the heat motion energy of molecules (0.9 kcal/mol at 37.5°C), then the dipole will be mainly at a fixed position.

For instance, the maximum electrostatic energy with which a carboxyl group binds to an -N^+H_3 cationic group in vacuum is −9.1 kcal/mol at $\mu = 2.8$ D, $\theta = 0°$, $d_e = 4.2$ Å (along an N-H bond).

Proteins do not have a diverse distribution or abundance of dipole groups in their peptide bonds, whereas drug molecules have a wider distribution and abundance of dipole groups. As the direction of the dipole moment of the C-H bond is unknown for many drugs, we assign μ values for opposite polarity directions in the C-H bond (C^-–H^+ и C^+–H^-). Thus, $\mu = 2.8$ D for the C=O bond (as shown in Table 1.6) represents the average value of 3.12 D and 2.4 D.

We calculated the charges on the poles using the following equation:

$$q = \frac{\bar{\mu}}{l} \tag{1.11}$$

where $\bar{\mu}$ is the mean dipole moment, and l is the bond length.

The $\bar{\mu}$ value of most of the diatomic groups, which are often found in drug molecules, ranges from 1 to 2 D, except for the C = O and C ≡ N groups with greater dipole moments. Noteworthy, μ- and l–based calculations of charges on the poles of diatomic groups that are made disregarding atomic and homopolar dipoles and those of lone electron pairs are considered approximate values only.

The only experimental method by which the dipole moment of bonds can be calculated is by measuring the intensities of vibration absorption bands, but this approach does not enable the quantification of dipole moments of the entire molecule [13]. Thus, bond moments can be calculated from the latter values by applying vector composition laws. The dipole moment of a C-H bond depends on the type of carbon atom hybridization. Based on the infrared (IR) spectroscopy data, Petro [14] derived the following values of dipole moments for bonds:

$C^-_{sp^3}$ - H^+, 0.31 D; $C^-_{sp^2}$ - H^+, 0.63 D; C^-_{sp} - H^+, 1.05 D. The C-C bonds are also polar if they consist of atoms with a different type of carbon atom hybridization (Table 1.7).

Groups with high dipole moments may be diatomic or contributed from more atoms such as the amide group (Figure 1.13), with $\mu = 3.69$ D [15].

If a molecule has a large total quadrupole moment, then the ion-quadrupole interaction may play an important role in molecular recognition, although the interaction energy decreases when the distance is set as $1/d^3$. For example, quadrupole moments of benzene and heterocyclic compounds (pyrrole, furan, pyridine, and thiophene) have considerable values and constitute partly to the formation of cation-π complexes [16].

TABLE 1.6
Dipole Moments of Chemical Bonds [13] and Point Charges on Dipole Poles

Bond	μ, Debye				q,	Bond	μ, Debye				q, e⁻ charge
	C→H	C←H	$\bar{\mu}$	l, Å	e⁻ charge units		C→H	C←H	$\bar{\mu}$	l, Å	units
C – H	0.4	0.4	0	1.09	0.08	C – S	1.6	0.9	1.25	1.81	0.14
C – N	1.26	0.45	0.86	1.47	0.12	C = S		2.0	2.4	1.55	0.27
C = N		1.4	1.8	1.28	0.23	C – Se	1.5	0.7	1.1		
C ≡ N	3.94	3.1	3.5	1.15	0.64	H – O	1.51	1.51		0.96	0.33
C – O	1.9	0.7	1.3	1.42	0.19	H – N	1.31	1.31		1.01	0.27
(in alcohols)											
C – O	1.5	0.7	1.1	1.42	0.16	H – S	0.7	0.7		1.34	0.11
(in ethers)											
C = O	3.2	2.4	2.8	1.22	0.48	Si – H	1.0	1.0		1.48	0.14
C – F*⁾	2.19	1.39	1.8	1.36	0.27	Si – C		1.2	1.6	1.87	0.13
C – Cl*⁾	2.27	1.47	1.9	1.76	0.22	Si – N	1.55	1.55			
C – Br*⁾	2.22	1.42	1.8	1.93	0.20	N –:	~1.0	~1.0			
C – I*⁾	2.05	1.25	1.65	2.13	0.16						

*⁾ Dipole moments of the C-halogen atom bond in vinyl halogenides are ~0.7–0.8 D smaller.

TABLE 1.7
Dipole Moments of Carbon-Carbon Bonds [13, 14]

Bond	Bond Dipole Moment, D	Reference Molecule	Molecule Dipole Moment, D
$C^+_{sp^3} - C^-_{sp^2}$	0.69	Toluene	0.37
	0.67	Propylene	0.35
$C^+_{sp^3} - C^-_{sp}$	1.48	Propyne	0.75
$C^+_{sp^2} - C^-_{sp}$	1.15	Phenylacetylene	0.73

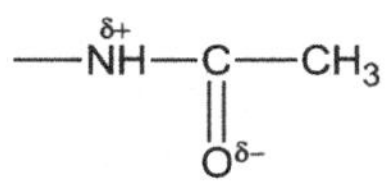

$$\overset{\delta+}{-NH}-\overset{\overset{\displaystyle ||}{C}}{\underset{O^{\delta-}}{}}-CH_3$$

FIGURE 1.13 Amide group.

1.7 DIPOLE–DIPOLE INTERACTION

If dipoles are arbitrarily positioned relative to each other and d >> l, then the energy of their interaction is

$$E = \frac{\mu_1 \mu_2}{d^3 \varepsilon} \cdot \left(2 \cos \alpha \cos \beta - \sin \alpha \sin \beta \right) \tag{1.12}$$

An explanation of the geometry parameters is provided in Figure 1.4. If one dipole is fixed, as seen in a protein molecule, and the other dipole is free, then the equation can be written as

$$E = \frac{\mu_1^2 \mu_2^2}{3k\ Td^6 \varepsilon^2} \cdot (1 + 3 \cos^2 \theta) \tag{1.13}$$

where k is the Boltzmann co nstant and θ is the angle of the fixed dipole axis with a line through the dipole centers.

As to the fixed relative position of the dipoles, when their interaction energy is greater than kT (espeically at distances near the dipole length, l), Equations (1.14) and (1.15), provided below, are applicable for the parallel and head-to-tail orientations of the dipoles, respectively:

$$E = \frac{\mu_1 \mu_2}{d^3 \varepsilon} \tag{1.14}$$

$$E = \frac{2 \mu_1 \mu_2}{d^3 \varepsilon} \tag{1.15}$$

If $d/l < 5$, a correction coefficient should be used in Equations (1.14) and (1.15), as the attraction energy is lower for the parallel orientation but higher for the head-to-tail relative position of the dipoles than its values determined by these equations; hence, the equation will take the form

$$E = A \frac{\mu_1 \mu_2}{d^3 \varepsilon} \tag{1.16}$$

Values of the correction coefficient A are presented in Table 1.8.

If μ is expressed in D and d is expressed in Å, then the potential energy E is

$$E = 14.4A \frac{\mu_1 \mu_2}{d^3 \varepsilon} \quad \text{kcal / mol} \tag{1.17}$$

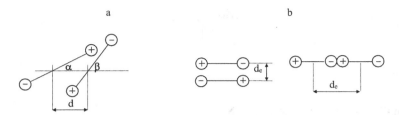

FIGURE 1.14 Scheme of dipole–dipole interaction: (a) arbitrary mutual position of the dipoles; (b) parallel and linear ("head-to-tail") positions with the maximum attraction energy.

TABLE 1.8
Values of the Correction Coefficient A for Different d/l [1]

d/l	Parallel Orientation	Head-to-Tail Orientation
1	0.586	—
2	0.845	2.67
3	0.914	2.25
4	0.962	2.13
>5	1	2

The equation for equilibrium dipole–dipole distance is written as

$$E = 10.8A \frac{\mu_1\mu_2}{d_e^3 \varepsilon} \text{ kcal / mol} \tag{1.18}$$

The formulas for dipole–dipole potential energy involve dielectric permittivity, which diminishes dipole interaction. However, this influence may be too weak, as dipoles cause a marginal orientation of water molecules. Various types of dipole groups exist in drugs, but only groups with 1 Debye or more contribute noticeably to the energy of complex formation with the receptor.

Dipole–dipole interaction energy depends on the contributions of all dipole groups of the molecule, all of which have a small total dipole moment at the same time. Nitrazepam (Figure 1.15) is a typical example of such molecule with sedative effects owing to specific binding with the benzodiazepine receptor in the brain.

Nitrazepam has a small total dipole moment characterized by its poor solvation in water, but the NO_2 and -C=O groups of the molecule have considerable dipoles.

For instance, to calculate the binding energy of two like diatomic dipoles with $\mu = 2$ D and $l = 1.5$ Å, the d_e value at the parallel orientation should be taken as 3 Å and that at the head-to-tail orientation as 4.5 Å ($\varepsilon = 1$); therefore, the corresponding E values are −1.35 kcal/mol and −1.07 kcal/mol.

The orientation of dipole groups in the molecules may differ from that of the dipole moments in the whole molecule.

Along with dipoles, multipoles of higher order, such as quadrupoles, octupoles, and others, may have molecular charges, but compared with dipole–dipole interactions, higher multipole–multipole interactions offer a much lesser contribution to complex formation energy. For example, the potential induced by a quadrupole and an octupole is proportional to $1/d^4$ and $1/d^5$, respectively.

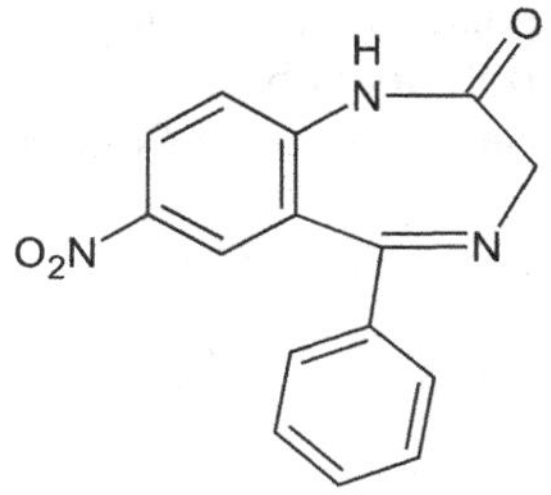

FIGURE 1.15 Nitrazepam.

1.8 HYDROGEN BONDING

A hydrogen bond is a three-center chemical bond formed between two electronegative atoms A and B, for example, $R_1AH\cdots BR_2$. The central hydrogen atom, which is covalently linked to atom A, forms an additional vdW bond with atom B, which has a lone electron pair directed along the line of this bond. A hydrogen bond can be either intermolecular or intramolecular. Electrostatic (dipole–dipole) interactions between the $R_1A^{\delta-} - H^{\delta+}$ and $B^{\delta-} - R_2^{\delta+}$ groups exchange repulsion energy at close distances, and the energy derived from the partial transfer of electron density from the BR_2 molecule to the R_1AH molecule is the main contributor to hydrogen bond energy. Hydrogen bond energy is maximum at the linear position of constituting atoms. The most stable hydrogen bonds are formed either when atom A and the substituents in the R_1AH molecule are mostly electronegative or when the A-H group can undergo considerable polarization under the influence of BR_2. Strong hydrogen bonds are formed by strongly electronegative atoms such as F, Cl, O, and N. Only the hydrogen atom bound to one of these atoms bears a considerable positive charge. The nitrogen atom is less electronegative than the oxygen atom; therefore, the energy of the O-H$\cdots$H hydrogen bond is greater than that of the N-H$\cdots$H bond. Hydrogen bond formation is associated with a decrease in the distance between atoms H and B (or between atoms A and B) as compared to the sum of vdW radii of the atoms engaged in hydrogen bond formation. For instance, the distance between the atoms in the H$\cdots$O bond in the linear water dimer is 2.04 Å, instead of 2.6 Å, and the distance between the oxygen atoms is 3 Å, instead of 3.6 Å.

The presence of a hydrogen bond in the IR spectra is displayed by a long-wavelength shift and broadening of the A-H band. The energy of hydrogen bond formation in liquid water is −4.5 kcal/mol, as derived from IR-spectroscopy measurements [1]. The energy of hydrogen bond formation in O-H$\cdots$O is within the range of 4–6 kcal/mol, whereas other biologically meaningful hydrogen bonds are more weak; for instance, hydrogen bond formation in N-H$\cdots$O is within the range of 2–3 kcal/mol and that in N-H$\cdots$N is within the range of 1–2 kcal/mol [1]. Although a proton, as a rule, remains in the R_1AH molecule, the latter, nonetheless, is conventionally referred to as the donor, and conversely, the BR_2 molecule acts as the proton acceptor. However, if groups that participate in hydrogen bond formation are strongly acidic or basic, then the intermolecular proton transfer may occur with the formation of a subsequent R_1A^- H$\cdots B^+R_2$ ion pair. In biological systems, ionic interactions are often accompanied by hydrogen bond formation, as for ammonium cation–carboxylate anion, and this combined interaction is commonly referred to as salt bridges.

1.9 ION-INDUCED DIPOLE INTERACTION

Overall, induced polarization of a molecule or its fragment in the ion field is a sum of electronic and atomic polarization. Atomic polarization accounts for 1%–5% of the total electronic polarization. Induced polarization chiefly involves the electrons at the valence atomic shell. Most molecules and bonds are asymmetric in terms of the polarizability: their longitudinal polarizability is greater than the lateral one. For example, the polarizability of benzene in the plane of the molecule is twice that of the polarizability in the cross direction. The energy of ion-induced dipole interaction is expressed by the following equation:

$$E = -\frac{\alpha_0 z^2 e^2}{2 d^4 \varepsilon^2} \tag{1.19}$$

where α_0 is the dipole static polarizability, and z is the ion valence.

If α_0 is expressed in units equal to 10^{-24} ml per one molecule, and if d is expressed in Å, then the potential energy of the interaction is calculated as

$$E = -166 \frac{\alpha_0 z^2}{d^4 \varepsilon^2} \text{ kcal / mol} \tag{1.20}$$

After correction for attractive forces that appear at equilibrium distances, the equation becomes

$$E = -111 \frac{\alpha_0 z^2}{d_e^4 \varepsilon^2} \text{ kcal / mol} \tag{1.21}$$

The polarizability is related to the molar refraction, the equation is expressed as $R_0 = \frac{4}{3} \pi N_A \alpha_0$, where N_A is the Avogadro's number. Values of polarizability for some bonds and small molecules are given in Table 1.10.

Refraction depends on the wavelength of the incident beam of light at which the measurement is made. By measuring refraction at two or more wavelengths, it is possible to calculate refraction at zero frequency (R_0) in a constant field. If R_0 is expressed in ml/mole, then the potential energy is calculated as

$$E = -\frac{3 R_0 z^2 e^2}{8 \pi N d^4 \varepsilon^2} = -65.8 \frac{R_0 z^2}{d^4 \varepsilon^2} \text{ kcal / mol} \tag{1.22}$$

and for the equilibrium distances:

$$E = -44.1 \frac{R_0 z^2}{d_e^4 \varepsilon^2} \text{ kcal / mol} \tag{1.23}$$

According to Webb [1], the interaction energy of a monovalent ion with a water molecule ($\alpha_0 = 1.444$) in vacuum at d = 5 Å due to dipole induction is estimated as −0.38 kcal/mol, which is much less than the interaction energy of the ion with the constant dipole of the water molecule (−5 kcal/mol). Interactions between ions and induced dipoles in the aqueous solution could be ignored in the case of polarization of relatively small groups. Nonetheless, induced polarization should be considered, especially if the ions interact with bulky groups such as the phenyl group ($R_0 = 25.11$ ml/mole). Thus, when a phenyl group interacts with an ammonium ion at the vdW equilibrium distance even in an aqueous medium, the contribution of the inductive component to the potential energy is approximately 0.15 kcal/mol.

1.10 DIPOLE-INDUCED DIPOLE INTERACTION

A permanent dipole can induce polarization of nonpolar molecules or bonds. Two permanent dipoles may further polarize each other. In such cases, the interaction energy is expressed by the following equation:

$$E = -5.71A \frac{R_0 \mu^2}{d^6 \varepsilon^2} \text{ kcal / mol} \tag{1.24}$$

where μ is the permanent dipole moment, and R_0 is the molar refraction of a nonpolar particle or polarizable dipole. Coefficient A equals either 2, if a polarized particle is in the dipole axis, or 0.5, if the particle lies on a line that is perpendicular to the dipole and that passes through the dipole center. Moreover, A = 1 for freely rotating molecules, and A = 2 when two like dipoles are mutually polarized [1].

For the vdW equilibrium distance, Equation (1.24) is expressed as

$$E = -2.85A \frac{R_0 \mu^2}{d_e^6 \varepsilon^2} \text{ kcal / mol} \tag{1.25}$$

A single dipole-induced dipole interaction does not play a significant role in the formation of the drug–receptor complex, but the sum of such interactions may contribute to its stability.

1.11 CHARGE TRANSFER COMPLEX

"Charge transfer" is a process wherein an electron of one molecule (donor) becomes partially bonded to another molecule (acceptor). The donor molecule must have a relatively low degree of electron (or electrons) localization, i.e., must have high-lying highest occupied molecular orbital. Charge transfer from one molecule to another occurs because of an overlap of electron shells of interacting molecules provided their electron affinities differ. The complexes formed due to charge transfer are subdivided into n- and π-complexes. In n-complexes, the acceptor orbital charge is transferred from the nonbonding orbitals of donor atoms (O, N, and S). As for model systems, wherein iodine is the acceptor and aliphatic amines, sulfides, esters, phosphines, and other compounds are donors, a significant increase in the dipole moments cannot be merely explained by the pure polarization effects. The acceptors that are weaker than iodine can also form charge transfer complexes. Thus, as assumed in [17], a minor increase (by 0.1–0.2 D) in the dipole moment occurs because of the charge transfer process during the formation of molecular complexes between the aforementioned donors and a weak charge acceptor such as carbon tetrachloride. Compounds without lone electron pairs but with an extensive π-system may act as donors.

Dipole moments of sandwich π-complexes formed by nonpolar partners are presented in Table 1.9; values indicate charge transfer, wherein the charge is equivalent to approximately 3%–5% of the electron charge.

Some heteroatoms or substituting groups in molecules or moieties containing conjugated bonds may cause electron density deficiency (e.g., in nitrobenzene) or surplus (e.g., in aniline) of. Pairs of such molecules with different electron densities can form charge transfer complexes, while the value of charge transfer rapidly approaches zero with an increasing distance. Charge-transfer complexes that are formed in the model systems are easily recognizable in ultraviolet and IR spectra; thus, their IR spectra are similar to those of hydrogen bond complexes. However, it is difficult to differentiate contributions of charge transfer and polarization to intermolecular interaction energy.

TABLE 1.9
Dipole Moments (D) of π-Complexes in Carbon Tetrachloride [18]

Donor	Acceptor		
	Tetracyanoethylene	Chloranil	2,4,6-Trinitrobenzene
Durene	1.26	–	0.55
Hexamethylbenzene	1.35	1.0	0.87
Naphthalene	1.28	0.90	0.69
Stilbene	–	–	0.82

1.12   DISPERSION INTERACTIONS

Dispersion forces (known as London forces) occur as a result of quantum mechanical fluctuations in electron density in atoms or molecules. The subsequently formed short-lived (instant) dipole or multipole moment induces dipole or multipole moment in neighboring molecules. The energy required for attraction between such multipoles is the energy of their intermolecular dispersion interaction. Dispersion interactions are versatile and can occur in any molecule, being dominant for nonpolar and weakly polar molecules or groups.

To evaluate dispersion energy, the London equation (1.26) or its modification (1.27) is used:

$$E = \frac{3\alpha_1\alpha_2 (h\nu_i)_1 (h\nu_i)_2}{2d^6 \left[(h\nu_i)_1 + (h\nu_i)_2 \right]} \tag{1.26}$$

$$E = \frac{3\alpha_1\alpha_2 I_1 I_2}{2d^6 \left[I_1 + I_2 \right]} \tag{1.27}$$

where α is the static polarizability of the molecule, h is the Planck's constant, ν_i is the frequency of the electronic transition oscillator, and I_n is the primary ionization potential of nth molecule.

Pauling and Pressman [19] proposed a mean I value of 14 eV. If α is measured in units of 10^{-24} ml per molecule, R is measured in ml/mole, I is measured in kcal/mol, and d is measured in Å, then the equations, wherein the parameters are expressed in kcal/mol, are transformed to

$$E = -242 \frac{\alpha_1\alpha_2}{d^6} \text{ kcal / mol} \tag{1.28}$$

or through R:

$$E = -38.1 \frac{R_1 R_2}{d^6} \text{ kcal / mol} \tag{1.29}$$

Webb [1] proposed the use of a coefficient (0.5) for the vdW equilibrium distance. Therefore, Equations (1.28) and (1.29) can be transformed as follows:

$$E = -121 \frac{\alpha_1\alpha_2}{d_e^6} \text{ kcal / mol} \tag{1.30}$$

$$E = -19\frac{R_1 R_2}{d_e^6} \text{ kcal / mol} \tag{1.31}$$

Slater and Kirkwood modified the London equation by introducing the factor $\sqrt[4]{Z_1 Z_2}$, where Z_1 and Z_2 are the number of electrons in the valence shells of interacting molecules or groups, and the new value of the electrical oscillator frequency [20]. Hence, Equation (1.29) can be rewritten as

$$E = -37.4\sqrt[4]{Z_1 Z_2}\frac{R_1 R_2}{d^6} \text{ kcal / mol} \tag{1.32}$$

For vdW equilibrium distance, we have the equation

$$E = -18.7\sqrt[4]{Z_1 Z_2}\frac{R_1 R_2}{d_e^6} \text{ kcal / mol} \tag{1.33}$$

Based on these equations, it is supposed that the London equation yields the low boundary of the dispersion energy, and the Slater-Kirkwood equation characterizes its upper limit. Table 1.10 yields the parameters of interatomic bonds or some molecules necessary for calculating the dispersion energy.

TABLE 1.10
Polarizability Values for Interatomic Bonds, Groups, and Some Molecules [1]

Bond or Molecule	Polarizability α_0, 10^{-24} ml/Molecule	Molar Refraction R_0, ml/Mole	Bond or Molecule	Polarizability α_0, 10^{-24} ml/Molecule	Molar Refraction R_0, ml/Mole
C – C	0.475	1.20	CO_2	2.59	6.59
C – H	0.655	1.65	NO	1.69	4.31
C – O	0.559	1.41	H_2O	1.44	3.67
C – N	0.598	1.51	NH_3	2.14	5.45
C – S	1.75	4.43	CH_4	2.70	6.85
C – F	0.705	1.78	C_2H_6	4.33	10.99
C – Cl	2.53	6.38	C_6H_6	9.89	25.11
C –Br	3.64	9.19	$-CH_2-$	1.77	4.50
C – I	5.57	14.08	$-CH_3-$	2.42	6.15
N – N	0.602	1.52	$-CH_2CH_3-$	4.19	10.65
N – H	0.721	1.82	-OH	1.28	3.26
O – O	0.641	1.62	-SH	3.56	9.05
O – H	0.733	1.85	$-OCH_3$	3.06	7.77
S – S	2.93	7.41	$-NH_2$	2.03	5.15
S – H	1.83	4.62	$-NH_3^+$	2.74	6.97
C = C	1.59	4.02	$-NO_2$	2.53	6.42
C = C	2.31	5.83	-COOH	3.06	7.78
C = O	1.31	3.32	$-COO^-$	2.33	5.93
C = N	1.86	4.69	$-COCH_3$	4.20	10.67
N = O	1.32	3.34	-CHO	2.43	6.17
			-CN	2.32	5.89

Group refraction includes refraction of this group bond with the carbon atom.

As the dispersion energy depends, to a certain degree, on polarization anisotropy, a more or less advantageous mutual orientation of interacting molecules or groups occurs. The energy of dispersion interactions does not depend on the dielectric constant of the medium because water molecules do not have time to orient in alternating high-frequency electrical fields and thus do not weaken them.

The refraction of a molecule or an atomic group is expressed as the sum of refractions of single bonds.

Given the polarizability parameters, the energy required for interaction between a phenyl group with a protein's nonpolar domain comprising, for example, of methylene residues can be estimated. Taking $R_1 = 25.1$ ml/mole, $Z_1 = 30$, $r_1 = 1.85$ Å for half size of the phenyl group and $R_2 = 4.50$ ml/mole; $Z_2 = 8$; $r_2 \approx 2.0$ Å for a methylene group, d_e is 3.8 Å, and the energy of dispersion interaction E_{disp} calculated using Equation (1.33) amounts for -2.68 kcal/mol. If $d_e = 5$ Å, then $E_{disp} = -1.07$ kcal/mol as calculated using Equation (1.32). The calculations are applicable for a one-sided contact between the phenyl group and protein, provided that each hydrocarbon chain of the phenyl interacts with one methylene group of the protein. Thus, the dispersion energy of binding of the methylene and phenyl groups apart at equilibrium distance is -0.3 kcal/mol or -0.06 kcal/mol at a distance of 5 Å.

As the apparent dispersion forces appear in the contact area, the interaction counterpart of a methylene group is the -CH= part of the phenyl group, not the whole phenyl.

Large nonpolar and weakly polar groups such as phenyl, indole, phenothiazine, cyclopentyl, cyclohexyl, and others can be considered as "anchoring" groups that greatly increase the affinity between a substance and its receptor.

However, even very small changes in anchoring large groups might cause changes in the substance's pharmacological specificity. For example, promazine (Figure 1.16) and its analogs containing a phenothiazine group are neuroleptics, whereas imipramine (Figure 1.17) and its analogs containing dibenzoazepine, dibenzo cycloheptadiene, or imidobenzyl are antidepressants. No compounds have the tricyclic system, which exert both neuroleptic and antidepressant properties at the same time.

1.13 HYDROPHOBIC INTERACTIONS

Hydrophobic interactions occur among nonpolar molecules or their fragments in the aqueous medium. Incorporation of a nonpolar molecule into water leads to breakdown of hydrogen

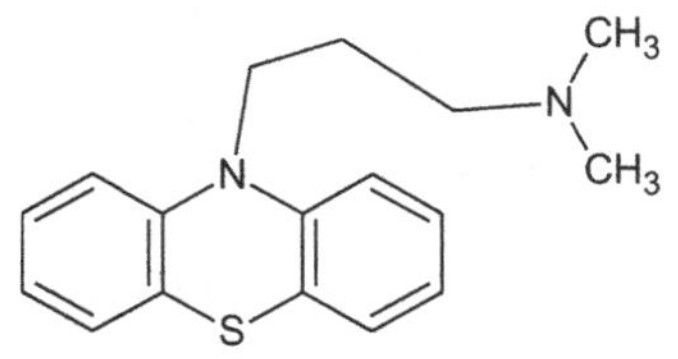

FIGURE 1.16 Promazine.

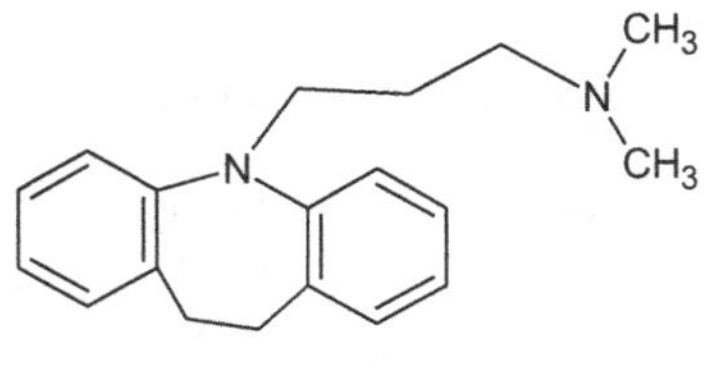

FIGURE 1.17 Imipramine.

bonds linking the water molecules. When compounds containing charged groups or atoms capable of forming hydrogen bonds are dissolved in an aqueous medium, the interaction between water molecules is superseded by the interaction of these atoms or groups with water. Such changes never occur if nonpolar molecules enter the aqueous phase: in this case, a balance in the thermodynamically disadvantageous contact with water is established through the association of hydrophobic molecules or their nonpolar fragments and disturbance in the water–water bonds. Hydrophobic interactions are always endothermic and proceed with increasing entropy. The entropic nature of the hydrophobic interactions causes strengthening of these interactions as the temperature increases. Hydrophobic interactions result from the association of many atoms or molecules and contribute to the formation of lipid micelles and bilayer membranes or the stabilization of supramolecular polymeric structures, such as tertiary and quaternary structures of protein macromolecules.

In our view, the role of hydrophobic interactions in the formation of complexes with low-molecular-weight ligands and proteins is often overestimated because of a correlation of the ligand binding free energy between proteins and ligands containing alkyl groups of various sizes to the corresponding increments in the free energy of alkyl group transfer from a water molecule to an organic solvent. However, this correlation may rather be putative. Values of the Hunch hydrophobicity constant π and $\Delta\Delta G$ of transfer of some alkyl groups from water to a nonpolar solvent and the dispersion energy of their one-sided interactions with proteins are shown in Table 1.11.

E_{disp} was calculated using Equation (1.33) as the energy required for binding to the surface of – CH_2– chains when d_e is 4 Å at the one-sided contact. Although these calculations yield approximate E_{disp} values, $\Delta\Delta G$ values used in the "extraction model" are comparable to those of E_{disp} produced at the submersion of the group into a protein-binding pocket.

1.14 COVALENT BONDING

A covalent bond is directly not involved in complex formation but rather participates in a chemical reaction of enzymes, receptors, and other macromolecules with drugs and biologically active compounds such as mercurial, arsenicals, antimonies, alkylating medicines used in cancer therapy, organophosphorus and carbamate acetylcholinesterase inhibitors, and other similar compounds. The functional activity of most of these drugs is based on the formation of short-lived complexes with binding energy of less than 15 kcal/mol in most cases. However, drugs whose action involves covalent bonding with binding energy of ~50–100 kcal/mol can irreversibly inactivate the corresponding biological target. Chemotherapy is the foremost and the most common field in which compounds

TABLE 1.11
Hansch Hydrophobicity Constants π [2] and $\Delta\Delta G$ Transfer Energy of a Number of Alkyl Group (in a n-Octanol–Water System) and the Calculated Energy of Dispersion Interaction with Proteins

	Octanol–Water		Dispersion Interaction Energy
Group	π	$-\Delta\Delta G$, kcal/mol	$-E_{disp}$ in One-Side Contact, kcal/mol
CH_2	0.50	0.68	0.26
CH_3	0.50	0.68	0.36
C_2H_5	1.00	1.37	0.59
C_3H_7	1.50	2.06	0.84
C_4H_9	2.00	2.74	1.16

with covalent bonding is being used. Some of these drugs include specific antibacterial and antiviral drugs as well as antitumor antibiotics.

As the covalent bonding process is preceded by a complex formation stage, gaining insight into the core mechanisms is important to design alkylating bioactive compounds and drugs with desired selectivity for specific molecular targets.

1.15 SHORT-RANGE REPULSIVE FORCES

If the intermolecular distance is around d_e, then nonspecific repulsive forces appear as a result of an overlap of electron clouds of the neighboring atoms.

The repulsion energy increases as the distance decreases proportionally to d^{-12}. The correction factor r for all types of interaction to consider repulsion at equilibrium distance is expressed as

$$r = 1 - \frac{b}{a} \tag{1.34}$$

where a and b are the distance-dependent indices for attractive and repulsive forces, respectively.

Equations to calculate the energy of various intermolecular interactions and the corresponding correction factors, r, are summarized in Table 1.12. To apply these equations for distances $d > d_e$, a factor $1/r$ is needed.

1.16 DIELECTRIC CONSTANT

To accurately describe the intermolecular electrostatic interactions in an aqueous medium, it is necessary to precisely calculate the dielectric constant, denoted as ε. The dielectric constant indicates the number of times the force between charged particles in a medium would be fewer than the force between them in vacuum. The Coulomb forces are weakened because of screening charges by dielectric molecules. The magnitude of ε characterizes the polarization degree of environmental molecules induced by the fields of dissolved particles. The dielectric constant of the medium is a macroscopic value and is related to its polarizability α by $\varepsilon = 1 + 4\pi\rho_0 \alpha$, where ρ_0 is the number of particles per volume unit.

The electronic, atomic, and orientation polarizations are commonly distinguished based on certain unique features. In the Coulomb field, each atom in a molecule oscillating near the equilibrium position moves mainly toward the field direction, thus generating a dipole. This movement causes atomic polarization, constituting ~1% of the overall polarization. Displacement of electron shells under the influence of the electric field and the consequent increase (as with polar molecules) or appearance (as with nonpolar molecules) of a dipole moment contributes to electronic polarization. The overall polarization is contributed majorly by the preferential alignment of permanent solvent dipoles in the electrical field of solute molecules' charges. The macroscopic dielectric permeability (ε_0) of water at 25 °C is equal to 78; this value shows the reduction in interaction energy between charged particles located at long distances (>20 Å).

As for an aqueous solution of a salt, the macroscopic dielectric constant is calculated by the equation proposed in [21]:

$$\varepsilon = \varepsilon_0 + 2\,\delta c,$$

where ε_0 is the dielectric constant of pure water, c is the molar concentration of the salt, δ is the molar depression of the dielectric constant, which is of 5.5 M^{-1} and 5 M^{-1} for NaCl and KCl, respectively. Concentrations of inorganic salts in the blood are of the order 10^{-2} M (0.9 g/l).

TABLE 1.12

Equations to Calculate the Energy of Intermolecular Interactions at Equilibrium Distances d_e and $d > d_e$

Interaction type	- E, kcal/mol		r
	d	d_e	
Ion–ion	$332\dfrac{z_1 z_2}{d\varepsilon}$	$305\dfrac{z_1 z_2}{d_e\varepsilon}$	0.92
Ion–dipole	$69.1\dfrac{z\mu}{d^2\varepsilon}\cos\theta$	$57.4\dfrac{z\mu}{d_e^2\varepsilon}\cos\theta$	0.83
Dipole–dipole	$14.4A\dfrac{\mu_1\mu_2}{d^3\varepsilon}$	$10.8A\dfrac{\mu_1\mu_2}{d_e^3\varepsilon}$	0.75
Ion–induced dipole	$166\dfrac{\alpha_0 z^2}{d^4\varepsilon^2}$	$111\dfrac{\alpha_0 z^2}{d_e^4\varepsilon^2}$	0.67
	$65.8\dfrac{R_0 z^2}{d^4\varepsilon^2}$	$44.1\dfrac{R_0 z^2}{d_e^4\varepsilon^2}$	0.67
Dipole–induced dipole	$14.4A\dfrac{\alpha_0\mu^2}{d^6\varepsilon^2}$	$7.2A\dfrac{\alpha_0\mu^2}{d_e^6\varepsilon^2}$	0.50
	$5.71A\dfrac{R_0\mu^2}{d^6\varepsilon^2}$	$2.85A\dfrac{R_0\mu^2}{d_e^6\varepsilon^2}$	0.50
Dispersion	$1.5\dfrac{\alpha_1\alpha_2}{d^6}\dfrac{I_1 I_2}{I_1+I_2}$	$0.75\dfrac{\alpha_1\alpha_2}{d_e^6}\dfrac{I_1 I_2}{I_1+I_2}$	0.50
	$0.236\dfrac{R_1 R_2}{d^6}\dfrac{I_1 I_2}{I_1+I_2}$	$0.118\dfrac{R_1 R_2}{d_e^6}\dfrac{I_1 I_2}{I_1+I_2}$	0.50
	$238.6\sqrt[4]{Z_1 Z_2}\dfrac{\alpha_1\alpha_2}{d^6}$	$119\sqrt[4]{Z_1 Z_2}\dfrac{\alpha_1\alpha_2}{d_e^6}$	0.50
	$37.4\sqrt[4]{Z_1 Z_2}\dfrac{R_1 R_2}{d^6}$	$18.7\sqrt[4]{Z_1 Z_2}\dfrac{R_1 R_2}{d_e^6}$	0.50

μ is expressed in Debye; α is expressed in units of 10^{-24} ml/molecule; R is expressed in ml/mole; I is expressed in kcal/mol, and d_e is expressed in Å. Other designations are the same as mentioned in the text.

Numerous solid substances have low ε_0 values, for example, ice, 3.3; octadecanol, 3; polyethylene, 2.3. This is because the molecules are rigidly fixed in a crystal lattice and orientation polarization is absent, whereas the electronic polarization becomes predominant.

The macroscopic constant ε_0 can be determined accurately. It is, however, difficult to determine the microscopic (local) ε value at short distances less than dozens of angstroms: as charged particles approach each other, the effective ε value decreases.

Unfortunately, no theoretical or experimental basis exists to confirm the superiority and preference of one method over another for an accurate determination of microscopic dielectric constants for intermolecular interactions [22]. Indirect approaches can be used to estimate ε dependence on the inter-charge distance from experimental data. It is possible that the most convincing estimation of ε_0 values for closely located ionic groups relies on the use of protonation constants K_{a1} and K_{a2} of polymethylene diamines $H_2N(CH_2)_nNH_2$ (see Table 7.2 of Chapter 7). The difference in the protonation energy of two amino nitrogen atoms is equal to the electrostatic energy of the second proton repulsion, which depends on the N–N distance. The ε magnitude is calculated according to equation (1.5) for a range of interionic distances for varying lengths of polymethylene chains.

In addition to the aforementioned diamines, Schwarzenbach proposed a reliable equation to determine the interatomic distances using rigid cyclic amines: piperazine and triethylene diamine [23].

$$\varepsilon = 6d - 11; \; \varepsilon = 6d - 12 \tag{1.35}$$

This equation (Equation (1.35)) was produced by approximating Schwarzenbach data [23, 24]. Similar calculations of the ε values can be performed using the dissociation constants for aliphatic dicarboxylic acids (see Table 7.1 of Chapter 7).

To calculate the potential interaction energy for monovalent ions, researchers used different relationships between ε and distance d. Based on data for aliphatic dicarboxylic acids, Conway and coauthors [25] proposed Equation (1.36) to evaluate the ε value in the distance range of 7–10 Å. According to Noyes [26], the ε–distance relation is expressed by Equation (1.37). Murcko [22] prefers to use Equation (1.38), while Warshel and coauthors [27] use Equation (1.39).

We intend to point out that the calculated ε values obtained using Equations (1.35)–(1.39) do not significantly differ for short distances from single-charged ions:

$$\varepsilon = 6d - 7, \; [25] \tag{1.36}$$

$$\varepsilon = 1 + 1.378\,(d - 0.054), \; [26] \tag{1.37}$$

$$\varepsilon = 4d \; [22] \tag{1.38}$$

$$\varepsilon = 1 + 60\,(1 - e^{-0.1d}) \; [27] \tag{1.39}$$

When ionic groups come into direct contact, water dipoles surround the ions beyond the contact zone and weaken the binding of ions, thus favoring the dissolution of ionic crystals such as NaCl.

The strength of an electric field produced by charged particles ranges in the following manner (starting from maximum): "ion–ion" > "ion–dipole" >> "dipole–dipole" interactions. Correspondingly, the influence of the dielectric constant due to the presence of peripheral water molecules around the active site on the binding energy changes in the same order. As for dipoles that immediately come into contact with water molecules, the ε value is often taken as 1.

As was said before, the dielectric constant of the medium does not influence the dispersion interactions because alternating fields have too high frequency to orient the water molecules.

1.17 HYDRATION OF MOLECULES

In an aqueous solution, both drug and receptor molecules are always associated with water molecules. Given the asymmetric charge distribution, a water molecule acts as a dipole, leading to drug–receptor interactions involving ion–dipole, dipole–dipole, and dispersion interactions with water, as well as hydrogen bonding. The total potential energy of the drug–receptor complex formation is expressed by the following equation:

$$E = E_K + E_{K-W} + E_{W-W} \tag{1.40}$$

where E_K is the energy needed for immediate drug–receptor binding, E_{K-W} is the energy needed for the displacement of water molecules from the contact zone, and E_{W-W} is the energy needed for the interaction between the displaced water molecules; the first and third parameters (complex-formation promoting contributors) in this equation are negative values, whereas the second parameter is a positive value. The strongest association with water is typical for ionic groups. Among primary and secondary hydration shells, the hydration energy of single-charge ions is presumably majorly (95%) contributed by the primary shell. The strength of the hydration shell depends on the charge density of the ions. The ions Na^+, K^+, Ca^{2+}, and Cl^- are bound to ionic groups present in a protein or drug without undergoing loss of the primary hydration shell. If inorganic ions lose their hydration shells, then the binding of drugs with the ionic groups of the receptors would be greatly hindered. The lower strength of the hydration shells surrounding the drug and protein molecules (as compared with inorganic ions) is attributed to the large size of the ionic groups in drugs and proteins and, therefore, a lower charge density.

For example, the affinity of ions to nicotinic acetylcholine receptors increases as the ionic radius increases. As for Na^+ with an ionic radius of 0.98 Å, the K_d value equals to $1.51 \cdot 10^{-1}$ M; for K^+ with an ionic radius of 1.33 Å), the K_d value equals to $9.2 \cdot 10^{-2}$ M, for Cs^+ (1.65 Å).

Other ions surrounded by a stable hydrate shell are as follows: for tetramethylammonium with an ionic radius of 3.5 Å, the K_d value equals to $3.8 \cdot 10^{-2}$ M [28]; for halogen ions F^- and Cl^- with ionic radii of 1.29 Å and 1.81 Å, respectively, the K_d equals to $5.2 \cdot 10^{-4}$ M [29].

To describe the binding between the ionic groups of drugs and receptors from a thermodynamical perspective, it is necessary to determine the number of displaced water molecules, their orientation relative to the ionic groups, and their interaction with peripheral water molecules. It is particularly challenging to determine the contribution of displaced water molecules to dehydration energy. As the displaced water molecules are not energetically equivalent, water molecules whose poles of dipoles are located on the same line as that of the charge centers or at a closer position gain the strongest association with ionic groups.

The hydration energy of dipoles is lower than that of ions. Because of hydrogen bonding, the surrounding water molecules violate the optimal orientation of the associated water molecules relative to a hydrated dipole.

A strong hydration shell can impede the penetration of a drug into the central nervous system. The strength of hydration shells in drugs is reduced with increasing size of the amino group, in the following order: primary > secondary > tertiary amine.

For example, Parkinson's disease occurs due to the deficiency of dopamine (Figure 1.18) in the basal ganglia of the brain.

However, dopamine, which contains the protonated highly hydrophilic primary amino group, shows poor penetration through the BBB and is ineffective as an anti-parkinsonism agent. Instead of dopamine, it is attractive to use (-)–dihydroxyphenylalanine (L-DOPA, Figure 1.19),

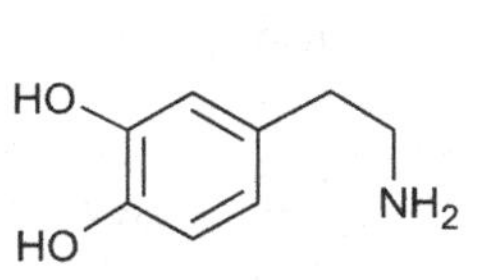

FIGURE 1.18 Dopamine.

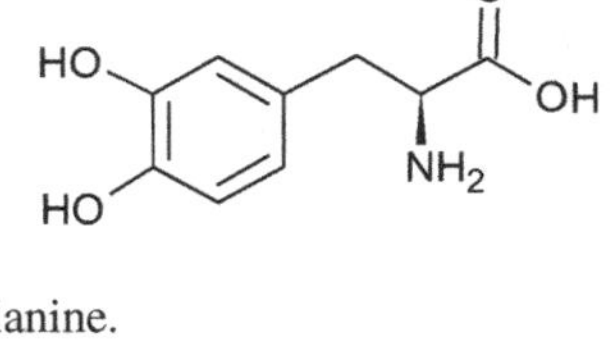

FIGURE 1.19 L-Dihydroxyphenylalanine.

especially through peroral administration, as it penetrates into the central nervous system and undergoes decarboxylation there with transformation to dopamine, thus causing the therapeutic effect [30].

The low hydrophilicity of the α-amino acid L-DOPA, as compared to dopamine, is due to the mutual neutralization of charges in the ionic groups.

Hydration effects in complex formation are discussed in more detail in Chapter 8.

1.18   RELATIONSHIP OF ENERGY COMPONENTS IN DIFFERENT TYPES OF INTERACTION

The total potential energy needed for the drug–receptor interaction is contributed by several components. It is important to know which of these components are vital and which of them can be neglected during calculations. Table 1.13 shows the calculated energies of different types of intermolecular interactions in water. All the calculations were made with the average equilibrium distance between monomers, d_e (4.5 Å). Dielectric constant was derived from Equation (1.35) for ion–ion interactions; half of ε was considered in case of ion–dipole interactions. To calculate the energy of dipole–dipole and dispersion interactions, the value of ε = 1 was used.

Although these calculated values are rather approximate, they indicate a proportion between the interaction energies of different types. In addition to the major types of interaction, dispersion interactions significantly contribute to the total energy, whereas a contribution from induced dipoles can be completely neglected. With increasing distance between a drug molecule and its receptor, the contributions of all interactions decline drastically, except for ion–ion interactions. At a distance of 8 Å, when one or two water molecules occupy the intermolecular space, the interaction energy of only monovalent ions (−1.12 kcal/mol) exceeds the energy of thermal motion.

TABLE 1.13
Energy of Different Intermolecular Interactions at a Distance of 4.5 Å between Monomers

Interaction Types	Ion–Ion		Ion–Dipole		Dipole–Dipole		Dispersion	
	-E,	%	-E,	%	-E,	%	-E,	%
	kcal/mol		kcal/mol		kcal/mol		kcal/mol	
Ion–ion	4.24	88.2	-	-	-	-	-	-
Ion–dipole	-	-	1.05	64.4	-	-	-	-
Ion–induced dipole	0.007	0.1	0.02	1.2	-	-	-	-
Dipole–dipole	-	-	-	-	1.04	71.7	-	-
Dipole–induced dipole	-	-	-	-	0.06	4.1	-	-
Dispersion	0.56	11.7	0.56	34.4	0.35	24.2	1.00	100
Total	4.81	100	1.63	100	1.45	100	1.00	100

* For an ion–ion dimer, $-N^+H(CH_3)_2 \cdots COO^-$ is used; the $-N^+H(CH_3)_2$ cationic group is used as an ion monomer; acetyl group is used as a dipole; phenyl group (benzene ring) is used to model dispersion interaction.

NOTE

1 Here and then, charges are calculated using Chimera software package [31], using the GAFF method.

REFERENCES

[1] J. Webb, Enzyme and Metabolic Inhibitors, vol. 1, New York: Academic Press, 1966, pp. 210–233.

[2] C. Hansch, J. Quinlan and G. Lawrence, *J. Org. Chem.*, vol. 33, p. 347, 1968.

[3] A. Albert, Selective Toxicity, London: Chapman and Hall, 1983.

[4] G. Kortum, W. Vogel and K. Anderssow, Dissociation Constants of Organic Acids in Aqueous Solution, New York: Plenum Press, 1961.

[5] A. Albert, Physical Methods in Heterocyclic Chemistry, vol. 1, Academic Press, 1963.

[6] H. Flaschka, A. Barnard and P. Sturock, Quantitative Analytical Chemistry, New York: Barnes & Noble, 1969.

[7] D. Perrin, Dissociation Constants of Organic Bases, New York: Plenum Press, 1965.

[8] R. Pearson and R. Dillon, *J. Am. Chem. Soc.*, vol. 75, p. 2439, 1953.

[9] L. Pauling, The Nature of the Chemical Bond, 3rd ed., Cornell University, 1960.

[10] R. Taft, Steric Effect in Organic Chemistry, New York: Wiley, 1956.

[11] L. Hammett, Physical Organic Chemistry, 2nd ed., New York: McGraw Hill, 1970.

[12] P. Kan, "Aliphatic amines," In: *Condensation Monomers*, J. Stille and T. Campbell, Eds., New Jersey: Wiley-Interscience, 1972.

[13] V. Minkin, O. Osipov and Y. Zhdanov, Dipole Moments in Organic Chemistry, Leningrad: Khimiya, 1968, pp. 73–77 [in Russian].

[14] A. Petro, *J. Am. Chem. Soc.*, vol. 80, p. 4230, 1958.

[15] C. Cantor and P. Schimmel, Biophysical Chemistry, vol. 1, New York: W. H. Freeman and Co., 1980, pp. 245–247.

[16] D. Dougherty, *Acc. Chem. Res.*, vol. 46, p. 885, 2013.

[17] S. Walker, Physical Methods in Heterocyclic Chemistry, vol. 1, New York: Academic Press, 1963.

[18] G. Briegleb, Elektronen Donator-Acceptor Komplexe, Berlin: Springer-Verlag, 1961.

[19] L. Pauling and D. Pressman, *J. Am. Chem. Soc.*, vol. 67, p. 1003, 1945.

[20] J. Slater and J. Kirkwood, *Phys. Rev.*, vol. 37, p. 682, 1931.

[21] J. Hasted, D. Ritson and C. Collie, *J. Chem. Phys.*, vol. 16, p. 1, 1948.

[22] A. Murcko, *J. Med. Chem.*, vol. 38, p. 4953, 1995.

[23] G. Schwarzenbach, *Pure Appl. Chem.*, vol. 35, p. 307, 1973.

[24] G. Schwarzenbach, *Z. Phyzik. Chem.*, vol. 176, p. 133, 1936.

[25] B. Conway, J. Bockris and I. Ammer, *Trans. Faraday Soc.*, vol. 47, p. 756, 1951.

[26] R. Noyes, *J. Am. Chem. Soc.*, vol. 84, p. 513, 1962.

[27] Y. Sham, Z. Chu and A. Warshel, *J. Phys. Chem. B.*, vol. 101, p. 4458, 1997.

[28] G. Akk and A. Auerbach, *Biophys. J.*, vol. 70, p. 2652, 1996.

[29] Y. Zhang, J. Chen and A. Auerbach, *J. Physiol.*, vol. 486, p. 189, 1995.

[30] M. Mashkovskii, Lekarstvennye sredstva (Drugs), vol. 1, Moscow, Novaya: Volna (in Russian), 2000, p. 140.

[31] E. Pettersen, T. Goddard, C. Huang, G. Couch, D. Greenblatt, E. Meng and T. Ferrin, *J. Comput. Chem.*, vol. 25, p. 1605, 2004.

2 Methods for Studying Drug-Receptor Binding

2.1 INTRODUCTION

The equilibrium and rate constants of drug–receptor binding along with the specific structural features of the ligands as well as the type of competition between substances provide the most important information to unravel the molecular mechanisms of complex formation. In most cases, the parameters of a receptor remain unknown. With the design of advanced computer programs, it becomes much easier to treat the experimental data for the determination of the rate and equilibrium constants [1]. Notwithstanding this, it is useful to present here the basic equations for the analysis and display of binding experimental data. A detailed description of the selected models, as well as ligand–receptor binding patterns not considered here, is provided in monographs [2, 3, 4].

Exploring the pattern of ligand–receptor binding based on the time and concentration of the interacting compounds, a certain binding scheme can be proposed. When comparing the predicted model and the experimental results, it is possible to exclude some suggested mechanisms of a ligand–receptor interaction. However, a good agreement of the model and the experiment is not always evident to favor a specific mechanism, as one may not rule out consistency of more than one kinetic scheme with the experimental results.

2.2 DETERMINATION OF EQUILIBRIUM CONSTANTS OF LIGAND–RECEPTOR BINDING

For the complex formation scheme

$$R + L \underset{k_{-1}}{\overset{k_{+1}}{\rightleftarrows}} LR \tag{2.1}$$

the following relationship holds true:

$$K_d = \frac{k_{-1}}{k_{+1}} = \frac{[L][R]}{[LR]} \tag{2.2}$$

which we can rewrite as

$$K_d = \frac{[L]([R_0] - [LR])}{[LR]} \tag{2.3}$$

DOI: 10.1201/9781003366669-3

where $[R_0]$ is the initial concentration of the free receptor. The latter relationship can be expressed as

$$\frac{[LR]}{[L]} = \frac{1}{K_d}[R_0] - \frac{1}{K_d}[LR] \qquad (2.4)$$

As the concentration of the ligand–receptor complex, $[LR]$, is assumed to be equal to the concentration of the bound ligand, the value of $[LR]/[L]$ is a concentration ratio of the bound ligand to the free one. A curve plotted in the coordinates ($[LR]/[L]$, $[LR]$) proposed by Scatchard [5] shows a straight line with a slope $-1/K_d$ and intercepting the abscissa axis at $[R_0]$ point, as depicted in Figure 2.1.

Obviously, plotting of the experimental concentrations $[LR]$, $[L]$ in the Scatchard coordinates makes it possible to derive the K_d (K_a) value and the concentration of the ligand binding sites. Deviations from the linear dependence in Scatchard coordinates point to a more advanced pattern of ligand–receptor binding than that depicted in Scheme (2.1).

Among possible reasons for the complexity of the ligand–receptor binding process, the most common are

- binding of a ligand to several acceptors,
- non–co-operative binding of a ligand to distinct sites of one receptor,
- co-operative binding of a ligand molecule with several identical or nonidentical sites, when the binding ability of these sites changes with the binding of each subsequent ligand molecule.

If a ligand molecule is bound to the independent sites of different types, then the curve plotted in Scatchard coordinates takes the form of a concave parabola.

When one ligand interacts with the binding sites of two distinct types, the parameters are determined in the following consecutive manner. To the areas of a curve, which correspond to the low and high concentrations of the added ligand, asymptotic lines are drawn (Figure 2.2). Values of the K_{d2} and K_{d1} equilibrium binding constants are determined from the slope of the corresponding asymptotes, while the concentration of the high-affinity sites $[R_{01}]$ and low-affinity sites $[R_{02}]$ is derived from the interception points of the corresponding asymptotes with the abscissa axis.

In case of cooperative binding, the curve in the Scatchard coordinates deviates from the linearity and has the upward or downward part depending on whether the cooperativity is positive or negative, respectively (Figure 2.3).

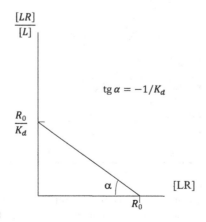

FIGURE 2.1 Determining the parameters of ligand–receptor binding using the Scatchard method.

Generally, the graphical representation of the experimental results derived from the Scatchard plot is one of the most informative methods to display binding data.

The curve for the binding of the ligand with the identical receptors under negative cooperativity has the same shape as that for the binding of the ligand with several non–co-operative receptors [6]. Bjerrum coordinates are used to discriminate these models of binding. Equation (2.3) can be transformed as follows:

$$[LR] = \frac{[R_0][L]}{K_d + [L]}$$ (2.5)

Hence, the extent of receptor saturation, y, is equal to

$$y = \frac{[LR]}{[R_0]} = \frac{[L]}{K_d + [L]}$$ (2.6)

The number of inflection points on a curve plotted with Bjerrum coordinates (y, log[L]) or (log y, log [L]) equals to the number of types of binding sites (Figure 2.4).

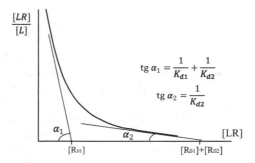

FIGURE 2.2 Determining the binding parameters using the Scatchard plot for an interaction between one ligand and two different binding sites.

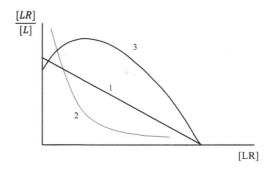

FIGURE 2.3 Plots in the Scatchard coordinates for different types of ligand-receptor binding. 1 - binding of a ligand with one receptor and no cooperativity; 2 - binding of a ligand with two types of non–cooperative sites and with one or more types of sites under negative cooperativity; 3 - binding under positive cooperativity.

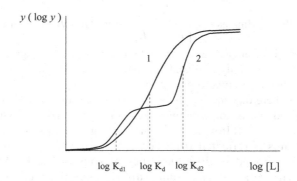

FIGURE 2.4   Bjerrum plots for ligand binding to one (1) or two (2) different receptor sites.

Cooperative ligand–receptor binding is revealed by plotting with the Hill coordinates [6], enabling to derive an index of cooperativity, n, which is characteristic of the binding capabilities of the receptor sites. The index of cooperativity varies as the ligand molecules consecutively bound to the binding site. Variations in the binding constants in this case are attributed to the fact that the receptor protein undergoes conformational changes within a time interval between the binding of preceding and subsequent ligand molecules. Generally, the cooperative binding can schematically be expressed by the following scheme:

$$L + R \overset{K_{d1}}{\rightleftarrows} LR; \quad L + LR \overset{K_{d2}}{\rightleftarrows} L_2 R; \; \ldots \; L_{n-1}R + L \overset{K_{dn}}{\rightleftarrows} L_n R \tag{2.7}$$

Under the assumption that the cooperative binding proceeds by the following pattern:

$$nL + R \underset{k_{-1}}{\overset{k_{+1}}{\rightleftarrows}} L_n R \tag{2.8}$$

the graphical analysis can substantially be simplified. Thus, analogously to Equation (2.3), one may write that

,
$$Kd = \frac{[L]^n \left([R_0] - [L_n R]\right)}{[L_n R]} \tag{2.9}$$

hence

$$\frac{[L_n R]}{[R_0] - [L_n R]} = \frac{[L]^n}{Kd} \tag{2.10},$$

and

$$\log \frac{[L_n R]}{[R_0] - [L_n R]} = n\log[L] - \log Kd \tag{2.11}.$$

A curve plotted with Hill coordinates $(\log\frac{[L_n R]}{[R_0]-[L_n R]}$, $\log[L])$ yields a straight line with a slope equal to n. The value of n is often used as a parameter of cooperativity degree. It is assumed that, if $n > 1$, then the extent of cooperativity of ligand–receptor binding is positive, being negative at $n < 1$; at $n = 1$, the binding is noncooperative (Figure 2.5).

It is worth mentioning here that similar plots in the Hill coordinates could represent either negative cooperativity or ligand interaction with two types of binding sites, with similar K_d values.

For example, consider the following case of binding analysis. If the ligand binding in the Scatchard coordinates is reflected from curve 2 in Figure 2.3, then it can be supposed that either the ligand is bound to different types of sites or the binding is characterized by negative cooperativity. Hill analysis cannot discriminate these cases, while the Bjerrum coordinates can be used for it. Whether we assume that the binding curve yields curve 1 as shown in Figure 2.4, then it is probable that the ligand is bound to identical sites under negative cooperativity.

Methodically, determination of the K_d is simplified when using labeled ligands. Competitive methods for determining the equilibrium and rate constants do not require labeling all compounds under study; it is sufficient to have one or very few labeled competitive ligands with the known binding constants. To determine the equilibrium binding constant of the unlabeled ligand K_d, it is convenient to use the Cheng–Prusoff equation [7]:

$$K_d = \frac{IC_{50}}{1+\dfrac{[L^*]}{Kd^*}}$$ (2.12),

where IC_{50} is the concentration of the unlabeled ligand at which binding of the labeled ligand is inhibited by 50%; [L*] and K_d* are the concentration and the dissociation constant of the labeled ligand, respectively.

2.3 METHODS FOR THE ANALYSIS OF LIGAND–RECEPTOR INTERACTIONS: TYPES OF ANTAGONISM

Methods considered in the previous section allow to estimate the type of binding sites and establish whether or not the binding of the same type of ligands to the receptors is cooperative.

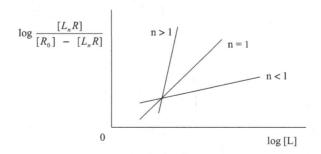

FIGURE 2.5 Determination of the extent of ligand-receptor binding cooperativity by plotting in the Hill coordinates.

In this section, we consider the methods for distinguishing between various types of antagonism to appear when two compounds bind to receptors of the same type. The major types of antagonism are

- competitive antagonism, when both ligands bind to the same receptor site
- noncompetitive antagonism, when two ligands interact with different sites of the same receptor
- uncompetitive antagonism, when a ligand of one type forms a complex with a receptor, and then a ligand of another type binds to this complex.

These types of antagonism can be discriminated by the method of double reciprocal coordinates, named after Lineweaver and Burk [8]. Equation (2.5) can be rewritten as

$$\frac{[L][R_0]}{[LR]} = Kd + [L]$$

Dividing both sides of the latter expression by $[L][R_0]$ gives

$$\frac{1}{[LR]} = \frac{Kd}{[R_0]} \cdot \frac{1}{[L]} + \frac{1}{[R_0]} \tag{2.13}$$

By plotting $1/[LR]$ against $1/[L]$, Eq. (2.13) gives a straight line with a slope equal to $K_d/[R_0]$ and intercepting the ordinate axis at $1/[R_0]$ and the abscissa axis, at $-1/K_d$.

In the case of binding of two ligands, L_1 and L_2, to receptor sites of the same type according to the pathway,

$$L_1 + R \underset{k_{-1}}{\overset{k_{+1}}{\rightleftarrows}} L_1 R \qquad L_2 + R \underset{k_{-2}}{\overset{k_{+2}}{\rightleftarrows}} L_2 R \tag{2.14}$$

the dissociation constants for the ligands are

$$Kd_1 = \frac{[L_1][R]}{[L_1 R]}; \; Kd_2 = \frac{[L_2][R]}{[L_2 R]}$$

$$[R_0] = [R] + [L_1 R] + [L_2 R].$$

Dividing both sides of the produced equation by $[L_1 R]$ yields

$$\frac{[R_0]}{[L_1 R]} = \frac{[R]}{[L_1 R]} + 1 + \frac{[L_2 R]}{[L_1 R]}.$$

Hence,

$$\frac{[R_0]}{[L_1 R]} = \frac{Kd_1}{[L_1]} + \frac{Kd_1[L_2]}{Kd_2[L_1]} + 1 \tag{2.15}$$

After dividing by $[R_0]$, we have the expression

$$\frac{1}{[L_1R]} = Kd_1\left(\frac{1}{[R_0]} + \frac{[L_2]}{Kd_2[R_0]}\right)\frac{1}{[L_1]} + \frac{1}{[R_0]} \tag{2.16}$$

Being plotted in $\left(\dfrac{1}{[L_1R]}, \dfrac{1}{[L_1]}\right)$ coordinates, Equation (2.16) would give a straight line with a slope

equal to $K_{d1}\left(\dfrac{1}{[R_0]} + \dfrac{[L_2]}{K_{d2}[R_0]}\right)$ and intercepting the ordinate axis at $1/[R_0]$ (Figure 2.6). With data

on the binding of one ligand in the absence and presence of its competitive partner, one can easily determine the equilibrium dissociation constants for both compounds of under study.

Similar transformations (not shown here) lead to the following double-reciprocal expression for noncompetitive binding:

$$\frac{1}{[L_1R]} = \frac{1}{[R_0]}\left(1 + \frac{[L_2]}{Kd_2}\right) + \frac{Kd_1}{[R_0]}\left(1 + \frac{[L_2]}{Kd_2}\right)\frac{1}{[L_1]} \tag{2.17}$$

The plots drawn for the binding of one and two noncompetitive antagonistic ligands cross the abscissa axis at the same point, $-1/K_{d1}$. Interception with the ordinate axis has a greater value for one ligand than for the noncompetitive binding of two ligands because, in the latter case, the number of binding sites is more than that in the former case (Figure 2.7).

In a case of uncompetitive antagonism of two compounds, the second ligand does not bind to a receptor but rather to a complex formed by the receptor and the first ligand. This case is written schematically as two types of interaction:

$$L_1 + R \underset{k_{-1}}{\overset{k_{+1}}{\rightleftarrows}} L_1R \qquad L_2 + L_1R \underset{k_{-2}}{\overset{k_{+2}}{\rightleftarrows}} L_2(L_1R)$$

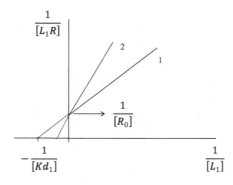

FIGURE 2.6   Lineweaver–Burk plots for binding of (1) one ligand and (2) two competitive ligands to a receptor.

To obtain the plots of experimental points with the Lineweaver–Burk coordinates, it is convenient to use the following equation:

$$\frac{1}{[L_1R]} = \frac{1}{[R_0]}\left(1 + \frac{[L_2]}{Kd_2}\right) + \frac{Kd_1}{[R_0]}\frac{1}{[L_1]} \tag{2.18}$$

Plotting of the data regarding the binding of one and two ligands gives two parallel straight lines with the same slope equal to $\dfrac{Kd_1}{[R_0]}$, but intercepting the ordinate axis at $\dfrac{Kd_1}{[R_0]}$ (line 1) or at $\dfrac{1}{[R_0]}\left(1 + \dfrac{[L_2]}{Kd_2}\right)$ (line 2), and the abscissa axis at $-\dfrac{1}{Kd_1}$ (line 1) or $-\dfrac{1 + \dfrac{[L_2]}{Kd_2}}{Kd_1}$ (line 2) (Figure 2.8).

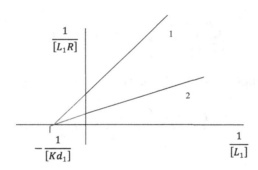

FIGURE 2.7 Lineweaver–Burk plots for binding of (1) one ligand and (2) two noncompetitive ligands to a receptor.

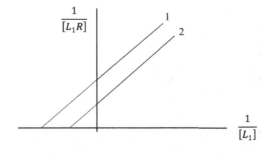

FIGURE 2.8 Lineweaver–Burk plots for binding of (1) one ligand and (2) two ligands in the case of uncompetitive antagonism to receptors.

2.4   PHARMACOLOGICAL METHODS FOR DETERMINING K_D AND THE ANTAGONISM TYPE

In most cases, experiments for determining the K_d values of unlabeled compounds are performed on isolated muscles of test animals. Distinguishing from radioligand methods, pharmacological methods rely on the use of unlabeled compounds. Muscle contraction or relaxation induced by the action of the substance as a result of binding to specific receptors is measured in the experiments. An extent of the effect (y) is assumed to be proportional to the degree of receptor saturation by the bound substance, i.e., $y = \dfrac{[LR]}{[R_0]}$. Values of K_d for antagonists are mostly determined by measurements of either muscle contraction at varying agonist concentrations in the presence of the antagonist at a constant concentration, or, conversely, under the effect of a fixed concentration of the agonist at varying concentrations of the antagonist. Examples of the plots are depicted in Figure 2.9.

As pharmacological methods do not allow determining the concentrations of receptor-associated and free ligand fractions, the measurements are carried out with the assumption that $[L] \approx [L_0]$, which holds true when $[L_0] \gg [R_0]$. Meanwhile, the concentration of the ligands must not be so excessive that they do not affect the metabolic processes in cells and, therefore, contractility of the muscles.

The curves shown in Figure 2.9 are described in Equations (2.19) and (2.21), which are easily produced by rewriting Equation (2.15) as

$$\frac{[R_0]}{[L_1R]} = \frac{1}{y} = \frac{Kd_1}{[L_1]}\left(1 + \frac{[L_2]}{Kd_2}\right) + 1,$$

where L_1 is an agonist. Multiplying both sides of the equation by $[L_1]$ gives

$$\frac{[L_1]}{y} = Kd_1\left(1 + \frac{[L_2]}{Kd_2}\right) + [L_1].$$

or

$$y = \frac{[L_1]}{Kd_1\left(1 + \dfrac{[L_2]}{Kd_2}\right) + [L_1]} \tag{2.19}$$

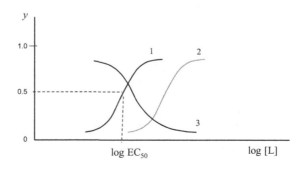

FIGURE 2.9   Plotting in coordinates (y, log [L]) for binding of the agonist (1) in the absence and (2) the presence of the antagonist at a constant concentration. Curve 3 is a plot in the coordinates (y, log [L_2]) at a constant concentration of the agonist. L_2 is an antagonist.

For occupying 50% of the receptors by antagonists (i.e., $y = 0.5$), we have

$$Kd_2 = \frac{\left[L_2\right]_{50}}{\dfrac{\left[L_1\right]}{Kd_1} - 1} \tag{2.20}$$

and in the absence of the antagonist

$$y = \frac{\left[L_1\right]}{Kd_1 + \left[L_1\right]} \tag{2.21}$$

If Clark's statement [9] holds true that the activity of the ligand is proportional to the number of occupied receptors. Taking the maximum effect of an antagonist as 100% and the observable effect, y, as the percentage of the maximum y_{max}, one can write the equation as

$$y = \frac{100\left[L_1\right]}{Kd_1 + \left[L_1\right]}$$

The curve plotted with the coordinates (y, $\log[L_1]$) has an S-shape with an almost linear part in the y ranging from 16% to 84%. In the case of competitive antagonism between the agonist and the antagonist, the corresponding linear parts on curves 1 and 2 (Figure 2.9) are parallel and have the same maximum.

The value of K_d for antagonists can also be determined using the Schild's method [10], which is widely used in pharmacological experiments to reveal the binding pattern of two competitively antagonistic ligands. If the agonist causes a certain effect at concentration $[L_1]$ alone and $x[L_1]$ in the presence of antagonist L_2, the equation may be written as

$$\frac{\left[L_1\right]}{Kd_1 + \left[L_1\right]} = \frac{x\left[L_1\right]}{Kd_1\left(1 + \dfrac{\left[L_2\right]}{Kd_2}\right) + x\left[L_1\right]}, \text{ hence}$$

$$x - 1 = \frac{\left[L_2\right]}{Kd_2}; \qquad \log(x-1) = pA_x - \log Kd_2 \tag{2.22}$$

where pA_x is a negative logarithm of the antagonist concentration.

Plotting of $\log(x-1)$ against pA_x (in the Schild coordinates) will give a straight line with a slope equal to 1 and an interception of pA_2 on the pA_x axis (Figure 2.10). Obviously, if $x = 2$, we have $pA_2 = \log K_{d2}$.

Deviation of the slope from unity may point to a noncompetitive interaction between agonists and antagonists, therewith $pA_2 \neq \log K_{d2}$.

With a graphical presentation of the experimental results by plotting in the double reciprocal coordinates$\left(\dfrac{1}{y}, \dfrac{1}{\left[L_1\right]}\right)$, it becomes possible to reveal the type of antagonism between the ligands at their binding to the receptors. Thus, in the case of competitive antagonism for two substances, accordingly to Equations (2.23) and (2.24), straight lines characterizing the binding of agonist L_1 in

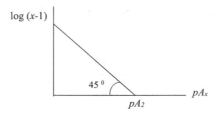

FIGURE 2.10 Plotting in the Schild coordinates.

the absence and presence of antagonist L_2 have the same interception point on the ordinate axis, with the same point having an ordinate 0.01.

$$\frac{1}{y} = \frac{1}{100} + \frac{Kd_1}{100} \cdot \frac{1}{[L_1]} \qquad (2.23)$$

$$\frac{1}{y} = \frac{1}{100} + \frac{Kd_1}{100}\left(1 + \frac{[L_2]}{Kd_2}\right) \cdot \frac{1}{[L_1]} \qquad (2.24)$$

Equations (2.23) and (2.24) are, respectively, analogous to Equations (2.13) and (2.16) with the expression of y as the percentage of the maximum effect (y_{max}). A graphical representation for the competitive antagonism is similar to that plotted in Figure 2.6, excluding the interception point on the ordinate axis, $1/[R_0]$, which becomes equal to 0.01. Under noncompetitive and uncompetitive antagonistic interactions, graphs in the coordinates ($1/y$, $1/[L_1]$) have a similar representation as depicted in Figures 2.7 and 2.8.

2.5   COMPARISON OF RADIOLIGAND AND PHARMACOLOGICAL METHODS OF K_D DETERMINATION

There is a widespread opinion that only radioligand methods provide a correct estimation of K_d values, while the constants, determined through pharmacological experiments, are believed to be putative. Stephenson [11] and Ariëns et al. [12] suggested that a maximum response of a muscle may be attained at various and even very little extent of receptor occupancy, depending on the "internal activity" of a ligand. Hence, it may be expected that the K_d values for agonists obtained by the pharmacological methods would be less than those obtained through radioligand measurements. However, there are different points of view about the validity of the approaches and measurement accuracy of ligand–receptor binding parameters determined through pharmacological experiments.

Table 2.1 shows a comparison of the constant K_d assessed mainly in radioligand experiments in the brain tissues of rats and guinea pigs and pharmacological experiments using isolated smooth muscles of rats and guinea pigs. Although Table 2.1 could be extended by the inclusion of ligands specific to other receptors, this would not interfere the conclusions that, first, the differences in K_d values determined for agonists by both methods are not always observed and, second, these constants, measured by different authors with the help of either method vary greatly, even for the same tissues. Because of the latter circumstance, it is impossible to more accurately compare the discussed methods for K_d determination. At least, the data presented in Table 2.1 give no reason to make conclusions that the pharmacological method is less precise than the radioligand one. The data presented in Table 2.1 show that it seems more important to standardize the procedures for

TABLE 2.1
Values of K_d Obtained Through Pharmacological and Rdioligand Methods

Ligand	Pharmacological Method			Radioligand Method		
	$K_d^{1)}$, M	Tissue	Refs.	K_d, M	Tissue	Refs.
Agonists:		mAChR$^{2)}$				
1. Acetylcholine	2×10^{-8}	Guinea pig intestine	[13]	4×10^{-8}	Rat brain	[17]
	$(4-6)\times10^{-8}$	—"—	[14]	8.3×10^{-8}	—"—	[18]
	1.5×10^{-7}	—"—	[15]	4×10^{-7}	—"—	[19]
	7×10^{-7}	guinea pig urinary bladder muscle	[16]	$(2-4)\times10^{-6}$	—"—	[20]
				$(2-4)\times10^{-6}$	guinea pig intestine	[20]
2. Muscarine	4×10^{-8}	Guinea pig intestine	[21]	1×10^{-8}	Rat brain	[22]
	2.5×10^{-7}	—"—	[13]	2.5×10^{-7}	—"—	[17]
	2.4×10^{-6}	guinea pig urinary bladder muscle	[16]	5×10^{-7}	—"—	[18]
3. Oxotremorine	2×10^{-8}	Rabbit seminal duct	[15]	2.4×10^{-9}	Rat brain	[22]
	4×10^{-8}	guinea pig intestine	[16]	7.1×10^{-9}	—"—	[20]
	4.4×10^{-6}	guinea pig urinary bladder muscle	[23]	8.7×10^{-8}	—"—	[19]
				$(5-8)\times10^{-7}$	—"—	[14]
				$(5-8)\times10^{-7}$	guinea pig intestine	[14]
Antagonists:						
4. Atropine	6×10^{-10}	Rabbit seminal duct	[24]	$(1-2)\times10^{-9}$	Rat brain	[14]
	1×10^{-9}	guinea pig intestine	[25]	$2,6\times10^{-9}$	—"—	[18]
	1×10^{-9}	—"—	[14]	2×10^{-10}	—"—	[19]
				$(2-4)\times10^{-9}$	guinea pig intestine	[14]
5. 3-QNB	3×10^{-11}	guinea pig intestine	[15]	6×10^{-11}	Rat brain	[14]
	5×10^{-10}	—"—	[14]	4×10^{-11}	human brain	[14]
	1×10^{-9}	guinea pig urinary bladder muscle	[16]	$(4-5)\times10^{-10}$	rat brain	[26]
				$(2-3)\times10^{-10}$	guinea pig intestine	[27]
1. N-methyl scopolamine (NMS)	1×10^{-9}	Rat intestine	[28]	7×10^{-10}	Rat intestine	[28]
				4.6×10^{-11}	rat brain	[20]
				2×10^{-10}	—"—	[29]

(continued)

TABLE 2.1 (Continued)
Values of K_d Obtained Through Pharmacological and Rdioligand Methods

Ligand	Pharmacological Method			Radioligand Method		
	$K_d^{1)}$, M	Tissue	Refs.	K_d, M	Tissue	Refs.
			β-Adrenoreceptors			
Agonists:						
1. Isoproterenol	2.5×10^{-9}	Guinea pig auricle	[30]	2.5×10^{-9}	Rat ventricle	[31]
	5×10^{-9}	guinea pig trachea	[30]	4.8×10^{-8}	rat brain	[32]
				1×10^{-7}	rat ventricle	[33]
2. Soterenol	1.9×10^{-8}	Guinea pig trachea	[34]	4×10^{-8}	Frog erythrocyte	[35]
	2.5×10^{-8}	"	[30]	2.4×10^{-6}	—	[36]
	7.9×10^{-8}	"	[30]	9.3×10^{-7}	Turkey erythrocyte	[37]
	5.6×10^{-7}	guinea pig auricle	[34]	5.3×10^{-7}	"	[37]
Antagonists:						
3. Propranolol	1.7×10^{-9}	Heart	[38]	5×10^{-10}	Guinea pig cerebral cortex	[39]
	2.5×10^{-9}	intestines	[38]	5.3×10^{-10}	Chinese hamster ovary cells	[40]
	2.5×10^{-8}	bronchi	[38]	8.6×10^{-9}	rat ventricle	[33]
	6.3×10^{-8}	guinea pig trachea	[23]	1×10^{-8}	frog ventricle	[33]
4. Practolol	4×10^{-8}	Heart	[38]	8×10^{-8}	Guinea pig cerebral cortex	[39]
	5.5×10^{-8}	intestines	[38]	5.1×10^{-7}	guinea pig lung	[41]
	1.6×10^{-7}	bronchi	[38]	1×10^{-6}	rat glioma	[42]
	7.9×10^{-6}	guinea pig trachea	[23]	8.6×10^{-9}	frog ventricle	[33]

1) It is a putative value when derived using the pharmacological method (see text).
2) Subtypes of mAChRs are not indicated, as the listed compounds possess no prominent selectivity for them [43, 44].

determining the activity of the ligands with respect to specific receptors and their subtypes. The advantages and limitations of both methods are listed in Table 2.2.

2.6 DETERMINATION OF THE RATE CONSTANTS OF LIGAND–RECEPTOR BINDING

If the ligand–receptor binding follows Scheme (2.1), then the change in the concentration of the complex formed between the interacting partners is expressed by the following equation:

$$\frac{d[LR]}{dt} = k_{+1}[R][L] - k_{-1}[LR]$$

with an assumption that $L \approx L_0$, $[L_0] \gg [R_0]$. As $[R] = [R_0] - [LR]$; therefore,

$$\frac{d[LR]}{dt} = k_{+1}([R_0] - [LR])[L_0] - k_{-1}[LR] = k_{+1}[R_0][L_0] - (k_{+1}[L_0] + k_{-1})[LR]$$

A solution of this equation is

TABLE 2.2
Advantages and Disadvantages of the Radioligand and Pharmacological Methods Used to Determine K_d

Advantages

(1) Clear and unambiguous theoretical background	(1) Simple experiment
(2) More information because of the possibility to determine the number of receptors in biological tissues	(2) The measured effect is due to ligand binding to specific receptors. This property is determined by the choice of a selective agonist
(3) Standard scheme of the experiment	(3) Enables the discrimination between agonists and antagonists
	(4) Relatively low cost of ligands

Disadvantages

(1) Separation of the bound and free labeled ligands may result in further dissociation of the bound ligand; redistribution of the label is one of the major sources of errors, especially for low- and medium-affinity ligands[1]	(1) Ambiguity of fundamental principles
(2) It is necessary to determine nonspecifically bound radiolabeled molecules; similar affinity of the ligands for various receptors can be a source of error	(2) Less information obtained; the method cannot determine the number of receptors in biological objects
(3) It is unclear whether the receptor state is sensitized or desensitized, differing by affinity. The relationship between these states depends on the incubation time of a tissue with the ligand	(3) Implicit error is possible in K_d value determination for agonists
(4) Minimal possibility to distinguish between agonists and antagonists or between the functional receptors and reserve receptors	
(5) Relatively high cost of labeled compounds	

[1] The handicap can be diminished with the use of high-affinity labeled ligands in combination with analyzed unlabeled compounds.

$$[LR] = \frac{k_{+1}[R_0][L_0]}{k_{+1}[L_0] + k_{-1}}\left\{1 - \exp\left[-t\left(k_{+1}[L_0] + k_{-1}\right)\right]\right\} \tag{2.25}$$

Dividing both the numerator and the denominator by k_{-1} produces the equation

$$[LR] = \frac{[R_0][L_0]}{[L_0] + Kd}\left\{1 - \exp\left[-t\left(k_{+1}[L_0] + k_{-1}\right)\right]\right\} \tag{2.26}$$

which can be rewritten as

$$[LR] = C\,[1 - \exp(-Kt)] \tag{2.27}$$

where

$$C = \frac{[R_0][L_0]}{[L_0] + Kd} \qquad \text{and } K = k{+}1\,[L0] + k{-}1 \tag{2.28}$$

By rearranging and taking logarithms, one can transform Equation (2.27) into the following expression:

$$\ln\left(1 - \frac{[LR]}{C}\right) = -Kt \tag{2.29}$$

which shows a straight line with a slope equal to $-K$ and passing through the origin. The procedure of the graphical determination of the k_{+1} and k_{-1} values includes two stages as represented in Figure 2.11.

1. The curve is plotted in semi-logarithmic coordinates $\{-\ln(1-[LR]/C),\ t\}$ for several $[L_0]$ values to determine the K values.
2. K is plotted against $[L_0]$, which gives a straight line with a slope k_{+1} according to Equation (2.28). The line crosses the ordinate axis at the point k_{-1}.

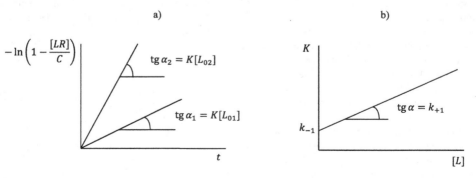

FIGURE 2.11 Graphical determination of ligand–receptor binding rate constants: (a) auxiliary K value; (b) k_{+1}, k_{-1}.

When the concentrations of the ligand and receptors are comparable, Equation (2.26) also provides a satisfactory description of the binding kinetics. If $[L_0] \ll [R_0]$, the rate constants can be evaluated with the modified equation, that is, Equation (2.27), where another expression for C and K is used. Thus, for this case [1],

$$C = \frac{[R_0][L_0]}{[R_0] + Kd},$$

$$K = k_{+1}[R_0] + k_{-1}$$

The rate constant of the formation of the drug–receptor complex, k_{+1}, ranges mostly from 10^6 to 10^8 $M^{-1}s^{-1}$. An experimental evaluation of [LR] at the binding stage is methodically difficult and laborious. Additionally, the relaxation methods for the analysis of fast enzyme–substrate or enzyme–inhibitor reactions are not always applicable for ligand–receptor binding. Thus, the k_{+1} value is often determined as $k_{+1} = \dfrac{k_{-1}}{K_d}$. Values of k_{-1} can be measured in experiments involving washing off of the receptors from the bound ligand. In the case of the dissociation of the ligand–receptor complexes through the pathway LR $\rightarrow$ L + R, one can consider $\dfrac{d[LR]}{dt} = -k_{-1}[LR]$ or $\dfrac{d[LR]}{[LR]} = -k_{-1}dt$.

Solving this equation gives

$$[LR] = [LR]_0 \exp(-t\,k_{-1}),$$

where $[LR]_0$ is the concentration of the ligand–receptor complexes at the start of the washing off. Transforming to the logarithmic form gives the equation

$$\ln\frac{[LR]}{[LR]_0} = -tk_{-1} \tag{2.29}$$

which, with the coordinates $(\ln[LR]/[LR]_0, -t)$, gives a straight line with a slope equal to k_{-1} (Figure 2.12).

The values of k_{-1} for the interaction of drugs with specific receptors are within a range of 10^{-3}–10^{-2} s^{-1}, corresponding to the half-life of the complexes $t_{1/2} = \dfrac{0,69}{k_{-1}}$, varying from 1 to 12 min.

2.7 COMPETITIVE METHODS TO DETERMINE THE RATE CONSTANTS OF LIGAND–RECEPTOR BINDING

Rate constants of an unlabeled ligand–receptor interaction can be determined by competition with labeled ligands. Motulsky and Mahan [45] described a simple and widespread model of a reversible competitive ligand and its competitor binding to the receptor. The model treats the reactions of the radioligand [L*] and the competitive inhibitor [I] with the receptor [R], resulting in the formation of receptor–ligand [RL*] and receptor–inhibitor (competitor) [RI] complexes, as depicted below:

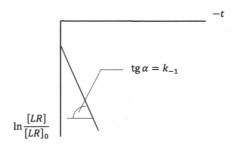

FIGURE 2.12 Determination of the k_{-1} value.

$$R + L^* \overset{k_{+1}}{\underset{k_{-1}}{\rightleftarrows}} RL^* \qquad R + I \overset{k_{+2}}{\underset{k_{-2}}{\rightleftarrows}} RI$$

where k_{+1} and k_{+2} are the rate constants for an association reaction of the receptor and the ligand and the inhibitor (competitor), correspondingly, and k_{-1} and k_{-2} are the rate constants for the reverse stage of the dissociation of the receptor and the ligand or the inhibitor (competitor) complexes. Hence, the equilibrium constant for the radioligand is expressed as $K_d = \dfrac{k_{-1}}{k_{+1}}$, and that for the inhibitor is expressed as $K_I = \dfrac{k_{-2}}{k_{+2}}$. For more convenient calculation, it is taken that no more than 10% of the radioligand and the inhibitor is involved in the binding. Hence, the concentrations of both free radioligand and the inhibitor slightly change as time progresses and are approximately equivalent to their initial concentrations. Therefore,

$$\frac{d\left[RL^*\right]}{dt} = k_{+1}\left[R\right]\left[L^*\right] - k_{-1}\left[RL^*\right]$$

$$\frac{d\left[RI\right]}{dt} = k_{+2}\left[R\right]\left[I\right] - k_{-2}\left[RI\right] \tag{2.31}$$

$$\left[R\right] = \left[R_0\right] - \left[RL\right] - \left[RI\right]$$

where [R] and [R_0] are the concentrations of the free and total number of receptors, respectively.

Solving the differential equation (Equation (2.31)) results in the expression for [RL*] as a function of time t:

$$\left[RL^*\right] = \frac{\left[R_0\right]k_{+1}\left[L^*\right]}{K_F - K_S}\left[\frac{k_{-2}\left(K_F - K_S\right)}{K_F K_S} + \frac{k_{-2} - K_F}{K_F}\exp\left(-K_F t\right) - \frac{k_{-2} - K_S}{K_S}\exp\left(-K_S t\right)\right] \tag{2.32}$$

containing new variables:

$$K_A = k_{+1}\left[L^*\right] + k_{-1}$$

$$K_B = k_{+2}\left[I\right] + k_{-2}$$

$$K_F = 0.5\left[K_A + K_B + \sqrt{\left(K_A - K_B\right)^2 + 4k_{+1}k_{+2}\left[L^*\right]\left[I\right]}\right]$$

$$K_S = 0.5\left[K_A + K_B - \sqrt{\left(K_A - K_B\right)^2 + 4k_{+1}k_{+2}\left[L^*\right]\left[I\right]}\right]$$

At $t = 0$, this equation reduces to zero. For the equilibrium state, two exponential terms in this equation are also approximating to zero and thereby can be omitted. Hence, the equation is simplified as follows:

$$[RL^*] = \frac{\left[R_0\right]k_{+1}k_{-2}\left[L^*\right]}{K_F K_S} = \frac{\left[R_0\right]\left[L^*\right]}{Kd\left(1 + \dfrac{[I]}{K_I} + \dfrac{[L^*]}{Kd}\right)} \tag{2.33}$$

In the absence of an inhibitor (competitor), the equation for radioligand binding is expressed as

$$[RL^*] = \frac{k_{+1}\left[R_0\right]\left[L^*\right]}{K_A}\left[1 - \exp\left(-K_A t\right)\right]$$

Equations (2.32) and (2.33) are used in numerical simulations of the competitive inhibition for a set of rate constants and concentrations [R], [L*], and [I]. The equations would differ when $k_{-2} \leq k_{-1}$ or $k_{-1} \leq k_{-2}$. Kinetic curves for radioligand binding in the presence or absence of the competitor are also different depending on how slowly or rapidly the release of the competitor (inhibitor) from the complex comprising the receptor occurs in comparison with the radioligand.

The use of the equation that describes the time dependence of a radioligand binding enables to calculate the association and dissociation rate constants of competing ligands for the obtained experimental dataset.

2.8 RELAXATION METHODS FOR DETERMINING THE RATE CONSTANTS

Direct kinetic analysis of the instantaneous stages of ligand–receptor binding, in the microseconds or seconds timescale, using the radioligand and pharmacological methods causes principal difficulties. First and foremost, direct observation of the system state is not possible because of the methods' inertial properties. Other limitations of these approaches are as follows:

(i) The radioligand method includes a relatively slow stage of separation of free and bound labels and requires the assessment of nonspecific binding extent

(ii) The pharmacological method offers an indirect evaluation of the binding parameters as judged by a reaction of the muscle in response to binding with ligands.

Therefore, both radioligand and pharmacological methods are applicable for determining the equilibrium constants of ligand–receptor binding, rather than the rate constants of the fast binding stage. The relaxation methods are commonly used to study the kinetics of instant processes such as chemical enzymatic reactions. Because of their high resolution, the relaxation methods can record the changes occurring within 100 ps.

A relaxation experiment involves the following stages:

1) A system under study is brought to equilibrium
2) The equilibrium state is disturbed abruptly to yield slight concentration changes when compared with the concentrations of the components constituting the system; the time of disturbance from the equilibrium state must therewith be shorter than the time of the registered processes
3) The time necessary to achieve a new equilibrium concentration of any component (or the relaxation time, τ) is recorded
4) Stages 1–3 are repeated for several concentrations of the components, and from the concentration dependency of the τ value, the kinetic scheme of the complex formation process or chemical reaction is determined.

A small instant shift of the equilibrium can be produced because of pulse changes in pressure and concentration, considering the known dependence of the equilibrium state of a chemical system upon these factors. Ranges of the rate constants for various molecular processes and the methods commonly applied for determining them are shown in Figure 2.13.

For example, consider the relaxation method of "temperature jump" (Figure 2.14). The temperature of the solution in a reactor increases to 10°C in few microseconds because of the discharge of a high-voltage capacitor. The temperature-related change is related to the characteristics of the capacitor and the reaction environment by

$$\Delta T = \frac{jCU^2}{2\rho C_\rho} \; ,$$

where j is the constant of the measuring cell, C is the capacity, U is the voltage, ρ is the density of the solution, and C_ρ is the heat capacity at constant pressure. The time of temperature change depends upon the capacity and the resistance of the reaction solution, $t = \frac{RC}{2}$.

An instantaneous heating of the system can be achieved through the use of microwave or laser pulse generators; often, a combination of "stopped flow" and "temperature jump" methods is successfully used for this.

To study ligand–protein systems through the methods of relaxation kinetics, the following conditions must be met:

1. Water solubility of the protein.
2. Presence of a high concentration of the investigated protein in the tissues.
3. A possibility to monitor the state of the ligand–protein system by changing the inherent fluorescence of the protein or optical density and spectral characteristics of the ligand molecule.
4. If measurements are made by the ligand, then it must be specific and contain fluorescence groups.
5. High purity of the protein, as nonspecific binding may complicate the kinetic analysis.

Being widely used for the investigations of chemical and enzymatic reactions, the relaxation methods, however, have yet found a limited application to study on the ligand–receptor binding kinetics because of the known difficulties to satisfy the given conditions.

Processes

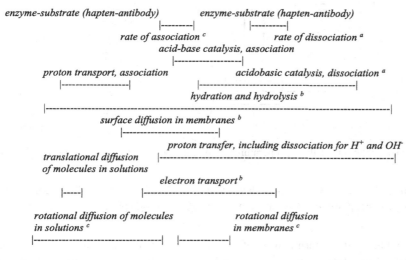

Methods

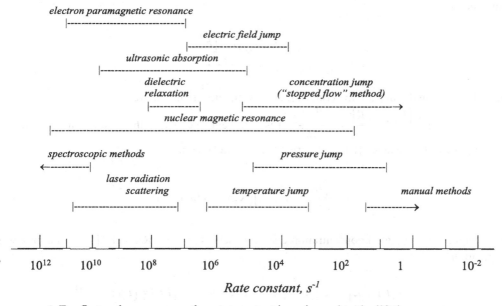

[a] For first-order processes, the rate constant is reciprocal to the lifetime.
[b] The rate constant is equal to that of second-order processes ($M^{-1}s^{-1}$).
[c] The rate constant is reciprocal to the time of rotational correlation.

FIGURE 2.13 Typical range of the rate constants of chemical and physical processes and range of rates accessible for measurement by different methods (based on data from [46]).

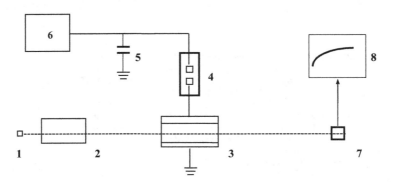

FIGURE 2.14 Schematic diagram of the "temperature jump" method based on the use of a high-voltage capacitor.

1 - light source; 2 - monochromator; 3 - reactor with solution under investigation; 4 - discharger; 5 - high-voltage capacitor; 6 - discharger power block; 7 - photomultiplier; 8 - oscillograph

The equations suitable for the determination of the rate constants through the relaxation kinetic methods can be derived [46]. Assuming that a little pulse change in the temperature to a new constant value led to a shift in the equilibrium of the system (Equation 2.1), the new equilibrium constant is equal to

$$Kd' = \frac{k'_{+1}}{k'_{-1}} = \frac{[L]'[R]'}{[LR]'} \qquad (2.34)$$

where [L]′, [R]′, and [LR]′ are the concentrations at the new temperature.

In as much as $\dfrac{d[L]}{dt} = \dfrac{d[R]}{dt} = -\dfrac{d[LR]}{dt}$, it is sufficient to consider the concentration change for one component; for instance [L], to describe the system behavior,

$$\frac{d[L]}{dt} = -k'_{+1}[L][R] + k'_{-1}[LR] \qquad (2.35)$$

With designations for small concentration changes such as Δ[L], Δ[R], and Δ[LR], one may write the equation as

$$\Delta[L] = [L] - [L]', \quad \Delta[R] = [R] - [R]', \quad \text{and} \quad \Delta[LR] = [LR] - [LR]' \qquad (2.36)$$

Substituting [L], [R], and [LR] from Equation (2.36) in Equation (2.35) gives

$$\frac{d([L]' + \Delta[L])}{dt} = \frac{d[L]'}{dt} + \frac{d\Delta[L]}{dt} = \frac{d\Delta[L]}{dt}$$
$$= -k'_{+1}([L]' + \Delta[L])([R]' + \Delta[R]) \qquad (2.37)$$
$$+ k'_{-1}([LR]' + \Delta[LR])$$

To derive Equation (2.37), we use that $[L]'$ does not change in time, that is, $\dfrac{d[L]'}{dt}=0$. As $\Delta[L] = \Delta[R] = -\Delta[LR]$, then

$$\begin{aligned}
\frac{d\Delta[L]}{dt} &= -k'_{+1}([L]' + \Delta[L])([R]' + \Delta[R]) + k'_{-1}([LR]' - \Delta[L]) \\
&= -k'_{+1}[L]'[R]' + k'_{-1}[LR]' \\
&\quad -\Delta[L]\{k'_{+1}([R]' + [L]' + \Delta[L]) + k'_{-1}\}.
\end{aligned}$$

(2.38)

Considering that $k'_{+1}[L]'[R]' = k'_{-1}[LR]'$ at the equilibrium state and $\Delta[L] \ll [L]'$, Equation (2.38) is reduced to

$$\frac{d\Delta[L]}{dt} = -\Delta[L]\left(k'_{+1}[R]' + k'_{+1}[L]' + k'_{-1}\right) = -\frac{\Delta[L]}{\tau}$$

(2.39)

$$\frac{1}{\tau} = k'_{+1}\left([R]' + [L]'\right) + k'_{-1}$$

(2.40)

where τ is the relaxation time.

An equation $\dfrac{d\Delta[L]}{dt} = -\dfrac{\Delta[L]}{\tau}$ has the following solution:

$$\ln\left(H[L]\right)_t = \ln\left(\Delta[L]\right)_{t=0} - \frac{t}{\tau}.$$

(2.41)

By plotting $\ln \Delta[L]$ against time past from a small and fast shift in the equilibrium, one may produce a straight line with a slope equal to $-\dfrac{1}{\tau}$ and intercepting the ordinate axis at $\ln (\Delta[L]_{t=0}$; Figure 2.15a).

According to Equation (2.40), plotting $-\dfrac{1}{\tau}$ against $([L]'+[R]')$ would give a straight line, having a slope of k'_{+1} and crossing the ordinate axis at k'_{-1} (Figure 2.15b).

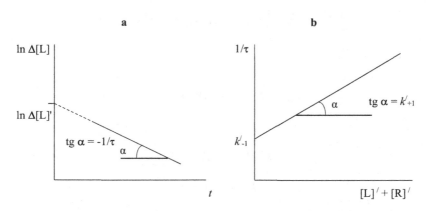

FIGURE 2.15 Determining the rate constants using methods of relaxation kinetics for one-stage processes.

Capabilities of the relaxation kinetic methods were first demonstrated in the fundamental studies of Hammes [47, 48]. The author investigated the interaction between aspartate aminotransferase with β-L-oxyaspartic acid and found seven intermediate states of the system, followed by 14 corresponding rate constants using the values of relaxation time. The detailed mechanism of the enzyme catalysis was then substantiated.

2.9   METHODS FOR THE STUDY OF CONFORMATIONAL KINETICS IN A LIGAND–RECEPTOR COMPLEX

Kinetics of conformational changes in a ligand–protein system are also studied using the methods of relaxation kinetics. As applied to the interaction described by the two-stage scheme

$$L + R \underset{k_{-1}}{\overset{k_{+1}}{\rightleftarrows}} LR \underset{k_{-3}}{\overset{k_{+2}}{\rightleftarrows}} LR^X \tag{2.42}$$

with the stage of conformational transformation of the primary complex LR into the form LR^X, the kinetic analysis resides in solving the following equations [49]:

$$\frac{d[R]}{dt} = -k_{+1}[L][R] + k_{-1}[LR] \tag{2.43}$$

$$\frac{d[LR]}{dt} = k_{+1}[L][R] - (k_{-1} + k_{+2})[LR] + k_{-2}[LR^X] \tag{2.44}$$

$$[R_0] = [R] + [LR] + [LR^X]$$

$$[L_0] = [L] + [LR] + [LR^X].$$

Assuming $\Delta[L] = \Delta[R]$, it can be written that $\Delta[LR^X] = -\Delta[R] - \Delta[LR]$; hence, Equation (2.44) can easily be rearranged into the form

$$\frac{d\Delta[R]}{dt} = a_{11}\Delta[R] + a_{12}\Delta[LR] + C_1$$

$$\frac{d\Delta[LR]}{dt} = a_{21}\Delta[R] + a_{22}\Delta[LR] + C_2 \tag{2.45}$$

where

$$a_{11} = -k_{+1}([L_e] + [R_e]);$$

$$a_{12} = k_{-1};$$

$$a_{21} = k_{+1}([L_e] + [R_e]) - k_{-2};$$

$$a_{22} = -(k_{-1} + k_{-2} + k_{+2}). \tag{2.46}$$

The subscript e indicates the equilibrium state. Linearization of Equations (2.43) and (2.44) is possible because if the small shift from the equilibrium state, that is, $\Delta[R] \ll [R_e]$, $\Delta[L] \ll [L_e]$, $\Delta[LR] \ll [LR_e]$, $\Delta[LR^x] \ll [LR_e^x]$.

Integrating the equation gives the equations with two exponential terms

$$\Delta[R] = C_{11}e^{\lambda_1 t} + C_{12}e^{\lambda_2 t} + C_1'$$

$$\Delta[LR] = C_{21}e^{\lambda_1 t} + C_{22}e^{\lambda_2 t} + C_2' \tag{2.47}$$

where C_{11}, C_{12}, C_{21}, C_{22}, C_1', and C_2' are constants, and λ_1 and λ_2 are the roots of the equation

$$\lambda^2 - (a_{11} + a_{22})\lambda + a_{11}a_{22} - a_{12}a_{21} \tag{2.48}$$

corresponding to the determinant constructed from the coefficients of the above equations:

$$\begin{vmatrix} a_{11} - \lambda & a_{12} \\ a_{21} & a_{22} - \lambda \end{vmatrix} = 0.$$

Roots of the quadratic equation (Equation (2.48)) are expressed as an irrational function:

$$\lambda_{1,2} = \frac{a_{11} + a_{22}}{2} \pm \sqrt{\left(\frac{a_{11} - a_{22}}{2}\right)^2 + a_{12}a_{21}}, \tag{2.49}$$

where "plus" and "minus" signs are related to λ_1 and λ_2, respectively. The relaxation times, τ_1 and τ_2, determined from experimental data are related to $\lambda_{1,2}$ by

$$\tau_{1,2} = -\frac{1}{\lambda_{1,2}} \tag{2.50}$$

Although the τ values are determined by the elementary constants, it is difficult to directly calculate the latter from relaxation times because of the irrationality of the function (Equation 2.50), except for the cases of significantly different relaxation times. Under experimental data handling, it is convenient to use equations connecting the values of τ_1^{-1} and τ_2^{-1}:

$$\tau_1^{-1} + \tau_2^{-1} = -(a_{11} + a_{22}) = k_{+1}([L_e] + [R_e]) + (k_{-1} + k_{-2} + k_{+2}) \tag{2.51}$$

$$\tau_1^{-1}\tau_2^{-1} = a_{11}a_{22} - a_{12}a_{21} = (k_{+1}k_{+2} + k_{+1}k_{-2})([L_e] + [R_e]) + k_{-1}k_{-2} \tag{2.52}$$

Plotting $\tau_1^{-1} + \tau_2^{-1}$ against a sum of the equilibrium concentrations of the ligand and the receptor, $[L_e] + [R_e]$ yields a curve with a slope equal to k_{+1} and an interception of the ordinate axis at the sum $k_{-1} + k_{+2} + k_{-2}$. The value of $k_{+1}k_{+2} + k_{+1}k_{-2}$ can be determined from a slope of dependence of $\tau_1^{-1} \cdot \tau_2^{-1}$ on the sum of the equilibrium concentrations $[L_e] + [R_e]$, while the $k_{-1}k_{-2}$ value can be determined from an intersection point on the ordinate axis.

REFERENCES

[1] H. Motulsky and R. Neubig, Current Protocols in Neuroscience, vol. 19, p. 7.5.1, 2002.

[2] L. Siegel, Ed., Biological Kinetics, Cambridge, England: Cambridge University Press, 1991.

[3] S. Varfolomeef and K. Gurevich, Biokinetics: A Practical Guide (in Russian), Moscow: FAIR-PRESS, 1999.

[4] N. Bindslev, Drug-Acceptor Interactions, London: Routledge, 2017.

[5] G. Scatchard, *Ann. N.Y. Acad. Sci.*, vol. 51, p. 660, 1949.

[6] A. Hill, *J. Physiol.*, vol. 40, p. 4, 1910.

[7] Y.-C. Cheng and W. Prusoff, *Biochem. Pharmacol.*, vol. 22, p. 3099, 1973.

[8] H. Lineweaver and D. Burk, *J. Am. Chem. Soc.*, vol. 56, p. 658, 1934.

[9] A. Clark, *J. Physiol.*, vol. 61, p. 547, 1926.

[10] H. Schild, *Br. J. Pharmacol.*, vol. 2, p. 251, 1947.

[11] R. Stephenson, *Br. J. Pharmacol.*, vol. 11, p. 379, 1956.

[12] E. Ariëns, J. Rossum and A. Simonis, *Arzneim. Forsch.*, vol. 6, p. 282, 1956.

[13] R. Clague, R. Eglen, A. Strachan and R. Witing, *Br. J. Pharmacol.*, vol. 86, p. 163, 1985.

[14] S. Snyder, K. Chang, M. Kuhar and H. Yamamura, *Fed. Proc.*, vol. 34, p. 1915, 1975.

[15] S. Shelkovnikov, *Advances in Science and Technics, Pharmacology Series*, 1991 (in Russian).

[16] L. Noronha-Blob, V. Lowe, A. Patton, B. Canning, D. Costello and W. Kinnier, *J. Pharmacol. Exp. Ther.*, vol. 249, p. 843, 1989.

[17] L.T. Potter, C.A. Ferrendelli, H.E. Hanchett, M.A. Hollifield and M.V. Lorenzi, *Mol. Pharmacol.*, vol. 35, p. 652, 1989.

[18] N. Birdsall, A. Burgen and E. Hulme, *Mol. Pharmacol.*, vol. 14, p. 723, 1978.

[19] H. Yamamura and S. Snyder, *Proc. Natl. Acad. Sci. USA*, vol. 71, p. 1725, 1974.

[20] M. Gillard, M. Waelbroeck and J. Christophe, *Mol. Pharmacol.*, vol. 32, p. 100, 1987.

[21] R. Blattner, H. Classen, H. Denhert and H. Doring, Experimente an Isolierten Glattmuskulären Organen, Freiburg: Hugo Sachs Elektronik KG, 1978.

[22] A. Closse, H. Bittiger, D. Langengger and A. Wanner, *Naunyn-Schmiedeberg's Arch. Pharmacol.*, vol. 335, p. 372, 1987.

[23] B. Levy and B. Wilkenfeld, *Eur. J. Pharmacol.*, vol. 11, p. 67, 1970.

[24] M. Eltze, *Eur. J. Pharmacol.*, vol. 151, p. 205, 1988.

[25] M. Eltze and V. Figala, *Eur. J. Pharmacol.*, vol. 158, p. 11, 1988.

[26] S. Snyder, G. Grenberg and H. Yamamura, *Arch. Gen. Psychiat.*, vol. 31, p. 58, 1974.

[27] S.-C. Lin, K. Olson, H. Okazaki and E. Richelson, *J. Neurochem.*, vol. 46, p. 274, 1986.

[28] L. Candell, S. Yun, L. Tran and F. Ehlert, *Mol. Pharmacol.*, vol. 38, p. 689, 1990.

[29] E. Hulme, N. Birdsall and N. Buckley, *Ann. Rev. Pharmacol. Toxicol.*, vol. 30, p. 633, 1990.

[30] J. Farmer, J. Kennedy, G. Levy and R. Marschall, *J. Pharm. Pharmacol.*, vol. 22, p. 61, 1970.

[31] T. Harden, B. Wolfe and P. Molinoff, *Mol. Pharmacol.,* vol. 12, p. 2, 1975.

[32] T. Cote, M. Munemura, R. Eskey and J. Kebabian, *Endocrinology*, vol. 107, p. 108, 1980.

[33] A. Hancock, A. De Lean and R. Lefkowitz, *Mol. Pharmacol.*, vol. 16, p. 3, 1979.

[34] G. Leclerc, B. Rouot, J. Velly and J. Schwartz, *Trends Pharmacol. Sci.*, vol. 2, p. 18, 1981.

[35] R. Kent, A. De Lean and R. Lefkowitz, *Mol. Pharmacol.*, vol. 17, p. 14, 1979.

[36] C. Mukherjee, M. Caron, D. Mullikin and R. Lefkowitz, *Mol. Pharmacol.*, vol. 12, p. 16, 1976.

[37] L. Brown, S. Fedak, C. Woodard, G. Aurbach and D. Rodbar, *J. Biol. Chem.*, vol. 251, p. 1239, 1976.

[38] A. Lands, A. Arnold, J. McAuliff, F. Luduena and T. Brown Jr., *Nature*, vol. 214, p. 597, 1967.

[39] D. Bylund, *Brain Res.*, vol. 152, p. 391, 1978.

[40] P. Pauwels, P. Van Gompel and J. Leysen, *Biochem. Pharmacol.*, vol. 42, p. 1683, 1991.

[41] G. Engel, D. Hoyer, R. Berthold and H. Wagner, *Naunyn-Schmeideberg's Arch. Pharmacol.*, vol. 317, p. 277, 1981.

[42] M. Magnize, R. Wiklund, H. Anderson and A. Gilman, *J. Biol. Chem.*, vol. 251, p. 1221, 1976.

[43] N. Buckley, T. Bonner, C. Buckley and M. Brann, *Mol. Pharmacol.*, vol. 30, p. 566, 1987.

[44] M. Waelbroeck, M. Tastenoy, J. Camus and J. Christophe, *Mol. Pharmacol.*, vol. 38, p. 267, 1990.

[45] H. Motulsky and L. Mahan, *Mol. Pharmacol.*, vol. 25, p. 2, 1984.

[46] A. Marshall, Biophysical Chemistry: Principles, Techniques and Applications, New York: Wiley, 1976.

[47] G. Hammes and J. Haslam, *Biochemistry*, vol. 7, p. 1519, 1969.

[48] G. Hammes and P. Schimmel, In: *The Enzymes*, vol. 2, P. Boyer, Ed., New York: Academic Press, 1970, p. 67.

[49] D. Koshland, *Fed. Proc.*, vol. 23, p. 719, 1964.

3 Methods for Drug Discovery Research

3.1 EXPERIMENTAL 3D STRUCTURE DETERMINATION OF DRUG–RECEPTOR COMPLEXES

Molecular recognition is based on structural and electronic features of the interacting molecules in essence. The more we know about effector and target structure details, the deeper is our understanding of molecular recognition determinants. The chemical structure of the molecules is routinely determined using many techniques, among which nuclear magnetic resonance (NMR) spectroscopy, X-ray crystallography, and cryo-electron microscopy (cryo-EM).

X-ray crystallography is based on scattering of X-rays on crystal lattice, and the resulting diffraction pattern can be decoded onto an electron density map of the molecule and then to its atomic structure. Cryo-EM is based on scattering of the electron beams on a frozen vitrified protein solution sample and detecting its electron images, followed by multi-step refinement and reconstruction of the molecular structure.

NMR spectroscopy uses nuclear spin–spin and spin–cell interactions in an external magnetic field. However, protein structure determination is much more difficult because of

- proteins function in cells, and biological liquids (plasma, intracellular liquid, and lipid membranes) are their typical environment; hence, it is difficult to isolate them
- proteins have a vast number of atoms within a single molecule; typically, biologically important enzymes and receptors have at least 200 residues and more than 1000 heavy atoms (except hydrogens), which creates difficulty in the high X-ray diffraction patterns resolution
- proteins have a tertiary structure and may form oligomers with highly organized multiple subunits.

Protein crystallography was one of the main directions of progress in molecular biology since Perutz and Kendrew (who received the 1962 Nobel Prize in Chemistry) resolved the relatively small and stable structure of myoglobin and hemoglobin using X-ray crystallography. As our main attention is focused on molecular recognition in neuroreceptors, it is worth noting that receptors directly forming ion channels (nicotinic acetylcholine receptors [nAChRs], glycine receptors) and receptors with different signaling pathways (mAChR, adrenergic receptors) are both transmembrane proteins. There have been specific difficulties in the neuroreceptor structure determination:

- instability of membrane-free proteins preventing the crystal sample from growing large enough for obtaining appropriately resolved X-ray diffraction images
- solubilization and purification of the protein in considerable quantities from the lipid membrane
- low quantities of signaling proteins in a tissue.

DOI: 10.1201/9781003366669-4

The first signaling protein for which the structure was successfully determined using cryo-EM was light-driven proton pump bacteriorhodopsin from bovine retina [1]. The bacteriorhodopsin has seven transmembrane helices containing extracellular and intracellular (cytoplasmic) loops, which is a common structural pattern of all G-protein–coupled receptors (GPCRs). Since then, gradual improvements in sample preparation, measurement, crystallization and purification techniques, and computer processing of experimental data for X-ray crystallography and cryo-EM have enabled the structural determination of α- and β-adrenergic receptors [2], mAChR [3, 4], nAChR [5], and many more neuroreceptors with a resolution of 1–2 Å, and understanding their signaling mechanisms. Because of the great progress in biomolecular structural studies, the 2012 Nobel Prize in Chemistry was awarded to Brian Kobilka and Robert Lefkowitz for GPCR studies, mainly by X-ray crystallography, while the 2017 Nobel Prize in Chemistry was awarded to Richard Henderson, Jacques Dubochet, and Joachim Frank for the development of Cryo-EM. Kurt Wüthrich was awarded much earlier (2002 Nobel Prize in Chemistry, partly awarded) for the application of NMR spectroscopy to studies on biomolecules.

The detailed knowledge of the receptor structure is a starting point for structure-based drug design [6]. When the receptor structure with the bound specific ligand is determined and the binding pocket is identified, it gives a possibility to virtually dock the predicted ligand present in the receptor pocket, analyze interacting groups, perform ligand binding free energy simulations, and accurately score the designed ligands. Details of molecular recognition affecting selectivity and affinity and identification of key interaction groups can be obtained from the receptor structure information. However, when only the apo-protein structure is known, it is challenging to determine the potential binding pockets. For example, the GTPase signal transducer KRAS protein, whose mutations play a significant role in cancer cell proliferation, has no well-defined binding pocket in an unbound state [7]. Such pocket is commonly named a "cryptic pocket" [8]. Nevertheless, the apo-protein structure aids in the investigation of protein dynamics and detection of the potential active sites. Recently, room-temperature X-ray crystallography with flash cooling has been developed, and it has allowed a detailed assessment of protein structural conformations [9, 10].

In addition to protein X-ray crystallography and cryo-EM methods, NMR spectroscopy, which is familiar to organic chemists, is also a very effective method for determining the structure of biomolecules [11]. Its main advantage is that it does not require a freezing protein sample and can be applied to proteins in the solution state. Thus, NMR spectroscopy is useful for investigating conformational changes in the protein structure. In recent years, NMR spectroscopy has been applied to the analysis of protein dynamics under near-physiological conditions [12]. A structural comparison between identical proteins determined using NMR spectroscopy or X-ray crystallography showed a significant difference in the crystal structure and NMR coordinates expressed in root mean square deviation (RMSD) [13]. The authors could not rule out the possibility of methodological issues and disorder factors in both X-ray and NMR coordinates. In a more recent paper, the authors proposed a Lindemann-like parameter for determining the fluctuation of individual protein residues in NMR spectroscopy data and constructed a model to consider the disorders in NMR protein coordinates, which may give an opportunity to identify disordered regions of a protein [14].

The features and advantages of the reviewed methods for the structure determination of protein–ligand complexes are briefly summarized in Table 3.1.

3.2 NEURAL NETWORK APPLICATIONS FOR PROTEIN STRUCTURE PREDICTION

The problem of protein folding is a question of how an amino acid sequence folds into tertiary and quaternary structures within a very short timescale, compared with the "regular" slow folding through the "trial and error" way [15]. The resolution to the problem is to directly connect protein structure reconstruction from the amino acid sequence. Stepwise progress in protein structure prediction using

TABLE 3.1
Comparison of Different Methods for Protein Structure Determination

Method	Features	Advantages
X-ray crystallography	Crystallization is required	High atomic resolution, up to 0.5 Å
	Relatively static protein structure	Highly developed techniques
	Room-temperature techniques available	Molecular weights ranging from small polypeptides to protein oligomers to DNA
Cryo-electron microscopy	Low resolution (barely less than 3 Å)	Short time for sample preparation
	Objects with 100 kD (membrane proteins)	Protein in the cell environment, no solubilization required
	Vitrification of the sample	
Nuclear magnetic resonance spectroscopy	Protein dynamics observation	Sample preparation is fast
	Purification is needed	High atomic resolutions
	Disorder factors for some protein chains	Protein in cell environment or in water solution

models, partly stimulated by the Protein Structure Initiative [16] and Critical Assessment of Protein Structure Prediction (CASP) competition, was culminated in 2021 and 2022, when the AlphaFold and AlphaFold2 neural network (NN) models by DeepMind were presented its unambiguously successful prediction with an RMSD of 0.96 Å for the 95th percentile of known proteins [17]. It is evident that we are on the brink of essential changes in protein structure prediction. Nevertheless, not an answer to the question: Why and how a drug is recognized by a receptor? But it was not intended to answer this question. Having such an effective predictive model, hypotheses on receptor protein states and possible ligand orientations can be easily verified *in silico* at a large scale [18, 19]. However, prediction continues to remain as a prediction with some probability. A targeted study on the structure prediction of transmembrane proteins (such as GPCRs) using AlphaFold2 indicated a well-performed model; yet, the authors recommend using the AlphaFold2 structural model cautiously [20].

3.3 "STRUCTURE–ACTIVITY" CORRELATION CONSTRUCTION

Increased activity and targeted action (selectivity) are often required for newly synthesized drugs. Examination of correlations between the characteristic of molecular structure and activity facilitates the elucidation of factors affecting activity and the proposal of hypotheses on the structure and functions of possible drug candidates. "Structure–activity" correlation models are frequently explored when the receptor structure is unknown, and ligand similarity features can be used for the prediction of their specific activity.

In the 1960s, Hansch and Fujita [21] as well as Free and Wilson [22] suggested a technique for the analysis of the quantitative structure–activity relationship (QSAR) for biologically active compounds. Hansch and Fujita [21] used two-parameter linear regression to determine the correlation between the minimal ligand concentration required for the manifestation of biological activity, as a dependent variable, and Hammet constant and hydrophobicity (octanol–water partition coefficient), as independent variables. Free and Wilson [22] suggested another variable—an indicator of the physico-chemical differences in the properties of a pair of molecules based on the topological characteristics of the molecular structure. Later, these variables were termed molecular descriptors. Other physico-chemical characteristics used by Hansch and Fujita as well as other authors for constructing correlation include polarizability, molar refractivity, pK_a, dipole moment, molecular weight, volume, and molecular surface area. Parameters of electronic structure such as ionization potential, charge of atoms, number of donor and acceptor atoms in the molecule, and the energy difference between the lower unoccupied molecular orbital and the highest occupied molecular orbital (HOMO).

The general scheme of QSAR methods includes construction of the statistically stable model for the "structural or electronic characteristic–activity" dependence [23]. The first step involves the collection of data for the tested set of compounds and the construction of one or another type of regression. In the second step, predictive power of the regression is examined occasionally multiple times. For this purpose, various statistical methods are employed for searching and determining parameters that provide the best correlation between the activity and structural characteristics for a massive multidimensional dataset. These methods can consider heterogeneity (presence of outliers) and possible interrelation of data. In particular, the principal component analysis searches for a linear combination of variables that identifies the maximally independent descriptors. The partial least squares method enables the selection of a limited set of significant variables (both dependent and independent) during iteration from a large number of initial dataset. If the paired cor-relation coefficient is nearly zero, then these two variables are considered independent. To eliminate spurious correlations, multiple cross-checks of regression that exclude one or several variables from the set are performed. Regression quality is evaluated as the square of the correlation coefficient. Meanwhile, the standard deviation of the mean should also be small during exclusion testing.

The values of the correlation coefficient and standard deviation of the predicted activity levels from the experimental data represent clear criteria of the quality of the QSAR model. It is worth mentioning that predicting experimental characteristics without any *a priori* suggestions of the factors affecting activity (or any other predicted property) is an attractive characteristic of the QSAR method. However, reliability of the prediction cannot always be evaluated in advance. It is mainly defined by the quality of the set of compounds (diversity of molecules in terms of activity and struc-tural characteristics).

Molecular descriptors are used for constructing QSAR that includes various topological indices to measure molecular similarity (such as Tanimoto coefficient), and the majority of the indices were introduced in the 1980s [24, 25].

The series of molecular connectivity indices χ^0, χ^1, χ^2, and so on was suggested by Kier and Hall [26]. The χ^0 index characterizes the degree of branching of the carbon skeleton of organic molecules (hereafter referred to as "skeleton"), that is, the sequence of atoms forming cycles and chains when all heteroatoms are replaced with carbon atoms and when the hydrogen atoms are ignored. The indices χ^1 and χ^2 (and so on) are determined by the number of two- and three-atom (and so on) fragments, respectively, and the number of bonds connecting these fragments to the neighboring atoms.

A so-called electro-topological state index (E-index), S_i, for atom i was introduced in [27] as a generalized measure of molecular similarity:

$$S_i = I_i + \Delta I_i$$

$$I_i = \frac{(2/n)^2 \chi^v + 1}{\chi^0}$$

$$\Delta I_i = \sum_{i \neq j} \frac{I_i - I_j}{r_{ij}^2},$$

where n is the principal quantum number of the atom valence shell, χ^v is the same connectivity index χ^0 that additionally considers bond multiplicity and lone pairs of electrons in the atom, r_{ij} is the atomic spacing, and I_i is the electronegativity index introduced by Kier. Successful qualitative classification of molecules from the database containing more than 20,000 different structures was performed based on the values of E-index, which resulted in organizing these compounds into structurally related classes.

Another descriptor defines the degree of molecule branching [25] associating each skeleton with a series of numbers $(p_0, p_1, p_2, \ldots)$, where element p_i equals the number of paths with the length of i bonds. Arranging the numeric series in descending order beginning from element 1, 2, and so on allows ranking of the structures according to their similarity (homology). The descriptor, defined not by the total length of the carbon chain in the molecule but by the number of cycles and the number of common atoms in them, was suggested in [28] as well as the heteroatom index that equals to the number of independent paths between heteroatoms, with the number of bonds (lengths) being 1, 2, and so on,. Good reproducibility of structure diversity and their clustering based on homology was demonstrated using ligands of various subtypes of muscarinic receptors as an example. Descriptors that display activity to a certain biotarget are found in the same region of the descriptor multidimensional space. Hash codes—which are lines with bit indices (0 and 1)—encoding molecules with pharmacologically significant structural features [29] are now widely used. Such codes allow programming effective comparison of the structures within a large dataset (several million entries).

In modern programs for the classification of chemical substances, the number of keys (descriptors) defining various aspects of the molecular structure, electronic structure, and physicochemical properties can range from several dozens to several thousands [30].

Structural indices do not describe the spatial shape of the molecule. Molecular similarity indices are used for this purpose (e.g., Tanimoto similarity index). Numerical characteristics of the shape of the molecular surface also correlate with activity, as the shape of the molecule defines steric complementarity mirroring the shape of the receptor active site under the molecule is in bioactive conformation [31, 32, 33].

3D QSAR methods are used for the evaluation of steric spatial factors facilitating activity [34]. Comparative molecular field analysis (CoMFA) is used for constructing "structure–property" correlation using a 3D pharmacophore [35]. This method allows comparing the effect of the distribution of the molecular electrostatic potential (MEP) in the vicinity of the pharmacophore groups in different molecules on their activity toward the target receptor. MEP is calculated for the set of points distributed in a cube of predefined size that involves all the test molecules with aligned pharmacophore elements. Linear regression is then obtained between the activity and the set of the MEP values at different areas on the molecule surface. CoMFA helps identifying spatial areas in the vicinity of the molecule that specifically participate in binding through intermolecular interactions, as well as assigns relative weights of each area to the activity using the expression for the energy of these interactions derived from QSAR analysis. It is often built using modern drug discovery software for the design of novel structures by comparing the similarity of the MEP regions in new chemical structures with the combination of the receptor's identified MEP regions.

Simple operation and high speed of data processing are the main advantages of QSAR methods. Unfortunately, very often, the introduction of small changes leads to the instability of the linear regression analysis. To construct a nonlinear correlation model, technologies of artificial NNs are actively employed [36, 37] (see also the previous section for other NN applications in drug discovery). This term is not related to real neurons but indicates the principle analogous to the operation of a neuron that transforms and transmits signals from one part of the system to another. The predictive model in NN technology is described by the number of neuron layers. Each standard neuron comprises a linear scalar function of input signal (vector of arguments), nonlinear transformation, and branching of the output scalar signal to other neurons. Neurons can be combined into layers, where all neurons of a layer receive one signal (see scheme in Figure 3.1). An NN is fully connected when all neurons transmit the output signal to all neurons including self.

Construction of the QSAR model is presented as optimization of the weight of neurons involved in signal transmission; therefore, a network is continuously changed (taught) until an acceptable error level is obtained for a training dataset. The trained NN is then used for prediction of

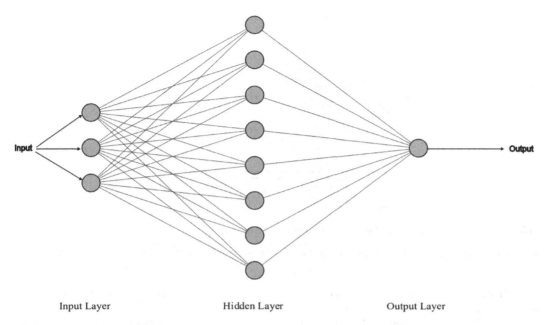

Input Layer   Hidden Layer   Output Layer

FIGURE 3.1 Simplified scheme of a layered neural network.

novel compounds' activity. The input may be the SMILES string of a compound's structure. For successful NN construction, consistent selection of molecular descriptors specific to the problem is very important. It can be achieved using machine learning–based algorithms.

Supervised machine learning methods for large datasets are used for ligand classification and scoring (active/inactive, strong binder/weak binder, etc.) in the drug discovery process. Among the methods, random forest and support vector machine algorithms are commonly applied [38].

3.4 BINDING AFFINITY MEASUREMENT AND HIGH-THROUGHPUT SCREENING METHODS

Experimental techniques for high-throughput binding affinity measurements are important for drug discovery, as they quickly yield information on the extent of drug activity on a target.

Surface plasmon resonance is a technique where polarized light refraction from a sensor conducting a thin-layer surface (typically gold) placed between the dielectric (e.g., glass) and the liquid sample is measured (Figure 3.2) [39]. Light emission can excite the electron plasmon wave present on a sensor's surface. For plasmon excitation (resonance), the light should have total internal reflection, the sensor layer should be smaller than the light wavelength, and the light frequency should be less than the frequency of a sensor's electron plasma waves (e.g., visible or near-infrared) [40]. Receptor protein molecules are immobilized on the functional coating of the sensor surface. Ligand binding is determined by changes in mass change during complex formation between the receptor and the ligand molecule (analyte), which leads to spectral and resonance angle changes. After washing off the analyte, the signal returns to baseline. With proper surface plasmon resonance (SPR) sensor calibration, binding concentration curves and the respective K_d as well as kinetic association and dissociation constants can be measured with high sensitivity.

SPR allows label-free (when compared with radiolabeling experiments described in Chapter 2) and direct analysis of ligand–receptor interactions. For immobilization of the receptor proteins, a proper coating of the sensor surface is applied for covalent or noncovalent attachment. The sensitivity

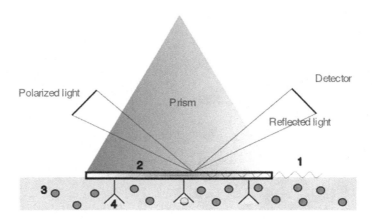

FIGURE 3.2   Scheme of the surface plasmon resonance experiment.

1 – Plasmon wave; 2 – Thin conducting layer; 3 – Analyte molecules; 4 – Receptor.

of SPR can be increased by using nanolayers of graphene on the sensor surface or using gold nanoparticles as transport material for ligand molecules. SPR is successfully applied to studies of GPCR binding with ligands. For example, in [41], adenosine receptor binding with ligands from a small fragment library was investigated. Plasmon waveguide resonance modification of SPR is applied for ligand binding studies on GPCR receptors immobilized on cell membrane fragments or model lipid membrane, with the target receptor being expressed on cells [42, 43]. Regular SPR instruments have capacity to process approximately 500 samples per day. Improvements in the SPR technique have enabled the simultaneous processing of hundreds of samples for application in high-throughput screening (HTS) [39].

Isothermal titration calorimetry (ITC) is the only method used for the direct measurement of binding enthalpy and free energy. The principle of the method is shown in Figure 3.3.

The sample cell contains a protein for binding measurement and has input flow of ligand molecules (titrant). Heat change due to ligand binding is detected by temperature change $"T$ between the sample and reference cells, while feedback power applied to the sample cell to minimize the temperature difference is changed in a way depending on the sign of the heat effect. Time-dependent feedback power produces a series of peaks, and the integration over each peak yields binding enthalpy estimate $"H$. The position of the inflection point on the titration curve $"H$ vs. ligand/receptor molar ratio gives binding enthalpy and equilibrium binding constant K_d. ITC measurement is a preferable way to determine binding enthalpy and perform affinity optimization for drug candidates [45]. However, ITC is a low throughput technique, and it would be appropriate to use it during early screening stages of drug discovery when the hits are reliably identified. Fortunately, increasing throughput for calorimetry is not impossible [46].

HTS methods were proposed in the 1980s; since then, they are actively developed both in industry and in academia. The main idea behind developing HTS methods was to automatize testing while minimizing time and maximizing processing speed for hits generation, where hit is some flag of activity of a molecule on a tested target protein. Initially, 96-well plastic plates were used to process simultaneously, which then transformed to 384-well plates, and currently, 1536-well plates are used for routine processing on equipment with the corresponding number of microchannels (syringes) for automated flow of liquid (e.g., ligand in a buffer). Standard preparation of assays includes expression of the cloned receptor to cell culture or proteins immobilized on specific coated wells.

HTS experimental techniques for large-scale testing of small molecules to protein targets are based on signal detection from wells. Signal can be produced by using a fluorescent-labeled ligand

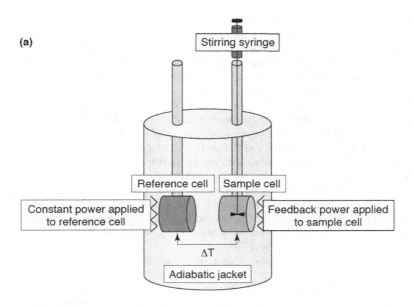

FIGURE 3.3   Scheme of an isothermal titration calorimeter.

(Source: Reproduced with permission from [44]).

or a radiolabeled ligand. After measuring the signal, statistical analysis of the generated datasets is performed. Good quality control and experiment planning is required, in addition to replication of the library screening experiment, to decrease false-positive and false-negative results. Selection of compounds for screening is also a complex task, and a large library is not always the best choice. For more details of HTS machinery and the current state of the methods, refer to previous literature [47, 48, 49].

The available methods for drug discovery are, by no means, limited to the reviewed methods. Our goal is to give the reader a broad perspective of the substantial progress made in this field in the past 30 years. However, molecular recognition principles can guide the drug discovery process even when the amount of available data is extraordinarily large. In the next chapters, some principles formulated in the Introduction section are explained and discussed thoroughly.

REFERENCES

[1] R. Henderson, J.M. Baldwin, T.A. Ceska, F. Zemlin, E. Beckmann and K.H. Downing, *J Mol. Biol.*, vol. 213, p. 899, 1990.

[2] V. Cherezov, D.M. Rosenbaum, M. Hanson, S. Rasmussen, F. Thian, T. Kobilka, H. Choi, P. Kuhn, W. Weis, B. Kobilka and R. Stevens, *Science*, vol. 318, p. 1258, 2007.

[3] T. Haga, *Proc. Jpn. Acad. Ser. B Phys. Biol. Sci.*, vol. 89, p. 226, 2013.

[4] A. Kruse, J. Hu, A. Pan, et al., *Nature*, vol. 482, p. 552, 2012.

[5] N. Unwin, *J. Mol. Biol.*, vol. 346, p. 967, 2005.

[6] M. Batool, B. Ahmad and S. Choi, *Int. J. Mol. Sci.*, vol. 20, p. 2783, 2019.

[7] J. Ostrem, U. Peters, M. Sos, J. Wells and K. Shokat, *Nature*, vol. 503, p. 548, 2013.

[8] S. Vajda, D. Beglov, A.E. Wakefield, M. Egbert and A. Whitty, *Curr. Opin. Chem. Biol.*, vol. 44, p. 1, 2018.

[9] J. Fraser, H. van den Bedem, A. Samelson, et al., *Proc. Natl. Acad. Sci. USA*, vol. 108, p. 16247, 2011.

[10] M. Maeki, S. Ito, R. Takeda, G. Ueno, A. Ishida, H. Tani, M. Yamamoto and M. Tokeshi, *Chem. Sci.*, vol. 11, p. 9072, 2020.

[11] K. Wüthrich, *Angew. Chem. Int. Ed.*, vol. 42, p. 3340, 2003.

[12] Y. Hu, K. Cheng, L. He, et al., *Anal. Chem.*, vol. 93, p. 1866, 2021.

[13] M. Andrec, D. Snyder, Z. Zhou, Z. Young, G. Montelione and R. Levy, *Proteins*, vol. 69, p. 449, 2007.

[14] E. Faraggi, K. Dunker, J. Sussman and A. Kloczkowski, *J. Biomol. Struct. Dyn.*, vol. 36, p. 2331, 2018.

[15] K. Dill, S. Ozkan, M. Shell and T. Weikl, *Annu. Rev. Biophys.*, vol. 37, p. 289, 2008.

[16] G. Montelione, *F1000 Biol. Rep.*, vol. 4, p. 7, 2012.

[17] M. Baek, F. DiMaio, I. Anishchenko, et al., *Science*, vol. 373, p. 871, 2021.

[18] G. Schneider and H.-I. Böhm, *Drug Discovery Today*, vol. 7, p. 64, 2002.

[19] B.J. Bender, S. Gahbauer, and A. Luttens, et al., *Nat. Protoc.*, vol. 16, p. 4799, 2021.

[20] T. Hegedús, M. Geisler, G. Lukács and B. Farkas, *Cell Mol. Life Sci.*, vol. 79, p. 73, 2022.

[21] C. Hansch and T. Fujita, *J. Am. Chem. Soc.*, vol. 86, p. 1616, 1964.

[22] S. Free and J. Wilson, *J. Med. Chem.*, vol. 7, p. 395, 1964.

[23] C. Breneman and M. Martinov, In: *Molecular Electrostatic Potential: Concepts and Applications*, J. Murray and K. Sen, Eds., Amsterdam: Elsevier, 1996, p. 149.

[24] R. King and D. Rouvray, Eds., Graph Theory and Topology in Chemistry, Amsterdam: Elsevier, 1987.

[25] M. Randič, In: *Concepts and Applications of Molecular Similarity*, M. Johnson and G. Maggiora, Eds., New York: Wiley Interscience, 1990, p. 77.

[26] L. Kier and L. Hall, Molecular Connectivity in Structure-Activity Analysis, Letchworth, U.K.: Wiley, 1986.

[27] L. Hall, L. Kier and B. Brown, *J. Chem. Inf. Comp. Sci.*, vol. 35, p. 1074, 1995.

[28] A. De Laet, J. Hehenkamp and R. Wife, *J. Heterocyclic Chem.*, vol. 37, p. 669, 2000.

[29] A. Hopfinger and B. Burke, In: *Concepts and Applications of Molecular Similarity*, M. Johnson and G. Maggiora, Eds., New York: Wiley, 1990, p. 173.

[30] Danishuddin and A.U. Khan, *Drug Discov. Today*, vol. 21, p. 1291, 2016.

[31] P. Dean, In: *Concepts and Applications of Molecular Similarity*, M. Johnson and G. Maggiora, Eds., New York: Wiley, 1990, p. 211.

[32] J. Grant and B. Pickup, *J. Phys. Chem.*, vol. 99, p. 3503, 1995.

[33] J. Grant, M. Gallardo and B. Pickup, *J. Comp. Chem.*, vol. 17, p. 1653, 1996.

[34] A. Hopfinger, S. Wang, J.J.B. Tokarski, M. Albuquerque, P. Madhav and C. Duraiswami, *J. Am. Chem. Soc.*, vol. 119, p. 10509, 1997.

[35] R. Clark, J. Leonard and A. Strizhev, In: *Pharmacophore: Perception, Development and Use in Drug Design*, O. Güner, Ed., La Jolla: IUL, 2000, p. 151.

[36] P. Liu and W. Long, *Int. J. Mol. Sci.*, vol. 10, p. 1978, 2009.

[37] I. Baskin, V. Palyulin and N. Zefirov, In: *Artificial Neural Networks: Methods and Applications*, D. Livingstone, Ed., USA: Humana Press, 2009, p. 133.

[38] L. Patel, T. Shukla, X. Huang, D. Ussery and S. Wang, *Molecules*, vol. 25, p. 5277, 2020.

[39] H. Nguyen, J. Park, S. Kang and M. Kim, *Sensors (Basel)*, vol. 15, p. 10481, 2015.

[40] D. Sotnikov, A. Zherdev and B. Dzantiev, *Biochemistry (Moscow)*, vol. 80, p. 1820, 2015.

[41] C. Shepherd, S. Robinson, A. Berizzi, et al., *ACS Med. Chem. Lett.*, vol. 13, p. 1172, 2022.

[42] I. Alves and S. Lecomte, *Acc. Chem. Res.*, vol. 52, p. 1059, 2019.

[43] E. Rascol, S. Villette, E. Harté and I. Alves, *Molecules*, vol. 26, p. 6442, 2021.

[44] G. Holdgate and W. Ward, *Drug Discovery Today*, vol. 10, no. 22, pp. 1543–1550, 2005.

[45] J. Ladbury, G. Klebe and E. Freire, *Nat. Rev. Drug Discov.*, vol. 9, p. 23, 2010.

[46] F. Torres, M. Recht, J. Coyle, R. Bruce and G. Williams, *Curr. Opin. Struct. Biol.*, vol. 20, p. 598, 2010.

[47] J. Hüser, Ed., High-Throughput Screening in Drug Discovery, Weinheim: Wiley-VCH, 2006.

[48] E. Yasi, N. Kruyer and P. Peralta-Yahya, *Curr. Opin. Biotech.*, vol. 64, p. 210, 2020.

[49] V. Blay, B. Tolani, S. Ho and M. Arkin, *Drug Discov. Today*, vol. 25, p. 1807, 2020.

4 Thermodynamics of Drug–Receptor Complex Formation

4.1 EQUILIBRIUM LIGAND–RECEPTOR COMPLEX FORMATION

Assuming that the receptor contains only one binding site of a ligand, complex formation between the ligand (L) and the receptor (R) can be expressed in the equation

$$L + R \underset{k_{-1}}{\overset{k_1}{\rightleftarrows}} LR \tag{4.1}$$

where k_1 and k_{-1} are the rate constants for a complex ligand–receptor association and dissociation, respectively. Reversible (equilibrium) processes are characterized by the equilibrium constant (K_e), which is also named as the association constant (K_a), with the M^{-1} dimension:

$$K_e = \frac{[LR]}{[L][R]} \tag{4.2}$$

More often, the inverse constant $K_d = \dfrac{1}{K_e}$, also known as the dissociation constant (measured in M), is used.

The constants K_d and K_e represent the complex stability measure. The lesser the K_d, the greater is the complex stability and ligand affinity. The Gibbs free energy change ΔG, which is used for the description of constant pressure and temperature processes, is also a measure of the ligand–receptor complex stability. Use of the constants K_d and K_e makes it possible to calculate ΔG under equilibrium conditions as

$$\Delta G = RT \ln K_d = 2.30\, RT \lg K_d \tag{4.3}$$

where R is the molar gas constant, and T is the absolute temperature. At 37.5 °C, which is the bath temperature in pharmacological assays involving isolated smooth muscle tissues, Equation (4.3) can be transformed into the following form:

$$\Delta G = 1.42 \lg K_d \text{ kcal / mol} \tag{4.4}$$

DOI: 10.1201/9781003366669-5

It follows that tenfold change in K_d results in 1.42 kcal/mol change in ΔG. When a ligand binds to a specific receptor, ΔG is a negative value, and the thermodynamic process is entirely exothermic. A ligand–receptor binding proceeds through different pathways and may involve several stages:

$$\begin{array}{c} \xrightleftharpoons[k_{-2}]{k_2} (RL)_2 \;\rightleftharpoons\; \cdots \;\xrightleftharpoons[k_{-m}]{k_m}\; (RL)_m \qquad (4.5) \end{array}$$

$$R + nL \underset{k_{-1}}{\overset{k_1}{\rightleftharpoons}} (RL)_1 \underset{k_{-2}}{\overset{k_2}{\rightleftharpoons}} (RL_2) \;\rightleftharpoons\; \cdots \;\underset{k_{-n}}{\overset{k_n}{\rightleftharpoons}}\; (RL_n) \qquad (4.6)$$

$$\xrightarrow{k_2} (R\text{-}L) \qquad (4.7)$$

- Pathway (4.5) involves transitions between distinct conformational states of a ligand–receptor complex.
- Pathway (4.6) pertains to the allosteric binding of several ligand molecules with a single receptor molecule beyond the orthosteric site. Cooperative allosteric binding means that attachment of each ligand molecule changes the receptor affinity to a next molecule. The cooperativity can be either positive if binding of a molecule increases the receptor's affinity or negative as in the case of decreasing affinity.
- Pathway (4.7) corresponds to the chemical bonding between a ligand and its target, a relatively rare event for pharmacodynamic agents[1].

The average kinetic energy during motion of molecules in solution (per mole) is 3/2 RT, that is, 0.89 kcal/mol at 25 °C. When $K_d = 10^{-1}$ M ($\Delta G = -1.36$ kcal/mol), the binding energy is slightly greater than the thermal motion energy, and the ligand–receptor complex is not stable. When the ligand and receptor in Equation (4.1) are present at a concentration of 10^{-4} M typical for biological systems, a tiny fraction of molecules (0.1%) exists in the form of complexes. If $K_d = 10^{-4}$ M ($\Delta G = -5.44$ kcal/mol), 38% of all molecules are in complex, while at $K_d = 10^{-7}$ M, the value is 97% ($\Delta G = -9.55$ kcal/mol; strong interaction) [1].

4.2 PARTIAL CONTRIBUTION OF DRUG MOLECULAR FRAGMENTS TO THE FREE ENERGY OF COMPLEX FORMATION

The regulation of the affinity of physiologically active compounds for specific receptors remains an important issue in molecular pharmacology. Low affinity of the drugs should be compensated by an increasing dose, but it can cause undesirable side effects. Conversely, too large affinity can result in high toxicity of a ligand at very low doses. For example, anticholinergic drugs have various degrees of affinity for mAChR with K_d values of 1.6 nM for atropine [2], 0.8 nM for scopolamine [3], 13 nM for pirenzepine [4], and 160 nM for adiphenine [5], respectively[2]. Single therapeutic doses of these drugs, at which the spasmolytic effect is exerted, are in good agreement with their affinities: for atropine and scopolamine, 0.25–0.5 mg (perorally and intracutaneously, respectively); for pirenzepine, 50 mg and 10 mg (perorally and intramuscularly, respectively); for adiphenine, 50–100 mg (perorally) [6].

The ΔG value directly related to affinity can be expressed as a sum of the following components:

$$\Delta G = \Delta G_E + \Delta G_W + \Delta G_{W-W} + \Delta G_{conf} + \Delta G_i \qquad (4.8)$$

where

- ΔG_E is pertaining to the free energy of ligand–receptor binding resulting from electrostatic and electrokinetic (exchange) forces
- ΔG_W is the desolvation free energy required to displace water molecules from a binding site
- ΔG_{W-W} is the free energy release during the association of the displaced water molecules

- $\Delta G_{conf.}$ is the free energy required for mutual adjustments in the conformation of the ligand and the receptor
- ΔG_i is the free energy required to displace the ionic atmosphere.

Because of its complexity, the ΔG value provides little information about drug–receptor binding, and it is expedient to use the method first substantiated by Pauling and Pressman [7] and extended by Webb [8a]. Consider two molecules, namely, A-B and A-C, where A is the main moiety, and B and C are relatively small groups with a similar pattern of interaction to the receptors. The value of $\Delta\Delta G = \Delta G_{A-B} - \Delta G_{A-C}$ can be attributed merely to a difference in the binding of the B and C groups. The other redundant and difficult factors can be neglected or cancelled out in that way. It is desirable that a substitution of one group for the other would not cause steric barriers for interaction. Calculations performed based on this approach were efficient for hapten-mediated suppression of antigen–antibody precipitation when using the constants of association between haptens and antibodies that are derived from the inhibition degree. Contributions of the substituents into the relative free energy, $\Delta\Delta G$, were compared with the calculated dispersion energies. The data [7, 9, 10] were somewhat modified by Webb [8b] with a correction for the number of electrons on the outer orbitals of the interacting groups. As seen in Table 4.1, the experimental $\Delta\Delta G$ values are in good correspondence with the calculated dispersion energies.

It is widely accepted that the relative binding free energy values offer extensive knowledge of the contribution made by different scaffold groups to receptor recognition. However, even the calculation of potential interaction energy (enthalpy) may predict the affinity of the ligands, as entropic contributions likely remain the same for a series of compounds with a similar structure.

4.3 CASE STUDY: mAChR ANTAGONISTS

Taking an experimental validation of the calculated energy data for more complex groups in ligand–receptor system as an example, we consider the binding of N-methyl-4-piperidyl esters of substituted glycolic acid to muscarinic acetylcholine receptors (mAChR). The ligands have a common structure in protonated form (Figure 4.1).

The experimental binding ΔG values between the selected ligands and mAChR obtained in rat small intestines are shown in Table 4.2.

Because of the high affinity for mAChR, an example of piperidyl esters of the substituted glycolic acid can be seen as suitable to search for quantitative relationships of complex formation with the receptor. Particularly, the $K_d = 3.4 \cdot 10^{-10}$ M of N-methyl-4-piperidyl benzilate [11] shows its high complementarity to the receptor.

TABLE 4.1

Comparison of Dispersion Interaction Energy between X-groups $X{-}\langle\!\!\!\bigcirc\!\!\!\rangle{-}AsO_3H^-$ of Haptens and Antibodies and Relative Binding Free Energy [9b]

Group X	Calculated energy E_{disp}, kcal/mol	Experimental value, $-\Delta\Delta G$, kcal/mol
CH_3	0.58	0.49
Cl	0.74	0.85
Br	0.89	0.89
I	1.05	1.10
OH	0.42	0.57
NH_2	0.52	0.29
NO_2	0.66	0.79
COO^-	0.53	0.52

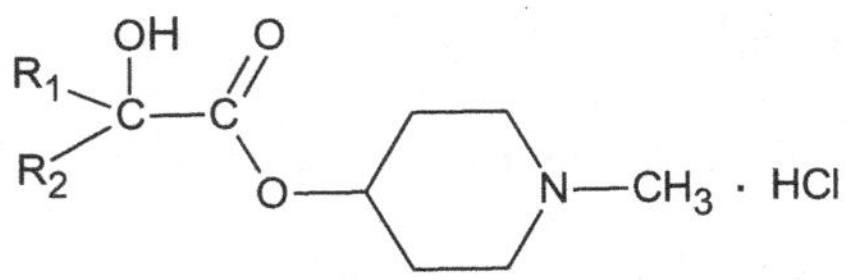

FIGURE 4.1   Structure of mAChR antagonists with the common N-methyl-4-piperidyl group.

Overall, the binding of the selected antagonists to their receptors is contributed by (1) the ion–ion interaction (salt bridge) between a protonated amino group and the receptor's anionic site; (2) dipole–dipole interactions between the oxygen-containing groups that can also form hydrogen bonds with the receptor; and (3) dispersion interactions, which are major for nonpolar or low-polar substituents. Because only nonpolar groups are varied in the series, they are expected to provide useful information about the dependence of dispersion interactions on the ligand's structure.

Using the known K_d values, it is possible to obtain fragmentary data regarding the contributions of chemical groups to the free energy of ligand–receptor complex formation. As with N,N-dimethyl-4-piperidyl esters of diphenylacetic and phenylacetic acids, the K_d values of the binding of these esters to rat brain mAChR are $1.0 \cdot 10^{-9}$ M and $4.6 \cdot 10^{-7}$ M, respectively [3]. Disregarding the contribution of substituting the hydrogen atom, we can estimate the $\Delta\Delta G$ value for a phenyl group as -3.7 kcal/mol. For the N-methyl-3-quinuclidinylic ester of diphenylacetic acid, a phenyl group has a less influence on the affinity for mAChR. As it follows from compared K_d for this compound and the same ester of phenylacetic acid ($1.2 \cdot 10^{-8}$ M and $1.4 \cdot 10^{-6}$ M, respectively), the $\Delta\Delta G$ value for the phenyl group is -2.9 kcal/mol. Methylation of a nitrogen atom in 3-quinuclidinyl benzilate reduces the affinity, whereas quaternization of atropine N-atom vice versa increases the affinity [3]. In the latter case, the contribution from a methyl group is 0.34–0.56 kcal/mol in magnitude.

A model to be used for calculations of the interaction energy of groups R_1 and R_2 with the receptor should consider the nature of forces engaged in anchoring of a group on the receptor as well as the steric conditions prevailing during the interaction, that is, whether the group is located on the protein surface or is immersed into the binding pocket, whether hindrances for binding occur because of the exceeded optimal size of the substituents. The model should involve the equilibrium distance between a ligand and the specific receptor. To verify this model, the calculated dispersion energy (E_{disp}) for groups R_1 and R_2 was compared with experimental relative free energy changes ($\Delta\Delta G$) of complex formation measured using the synthesized compounds [12, 13]. Partial contributions of $\Delta\Delta G$ for the substituents and the calculated dispersion interaction energies with the receptor are listed in Table 4.3.

Nonpolar or poorly polar groups R_1 and R_2 present in the compounds bind to the receptor protein mainly because of dispersion forces. The dispersion energy was calculated using Equation (1.33) without correction for the dehydration energy, which is relatively small for nonpolar groups and, in addition to this, can be compensated, to a certain degree, by the association of the displaced molecules. For dispersion parameters of the receptor protein, we used the following values: the number of electrons on valence orbitals, $Z = 8$; molar refraction, $R = 5.28$ ml/mol; and the radius of the groups, 2.0 Å. The use of these values was supported in [8, 9b] for calculating the dispersion energy of small molecules containing proteins. The used parameters are close to those for methylene and methyl substituents as well as some other groups that constitute the receptor's hydrophobic residues. As van der Waals radii of the substituents were used, the mean radii of the constituting atoms or groups were determined. In particular, the van der Waals radius of thienyl (1.82 Å) was calculated as the average radii (half-thickness) of a hydrocarbon chain (1.80 Å) and a sulfur atom (1.85 Å). Among different positions of substituents in aromatic cycles, we selected those positions at which the interaction energy is the maximum at the contact of the studied compounds with the receptor surface. The equilibrium distance, d_e, was considered as the sum of van der Waals radii for contact groups of the receptor (2.0 Å) and the substituent.

TABLE 4.2
mAChR Receptors Binding Free Energy of Compounds with the Common Structure Depicted in Figure 4.1 (Source: [12])

Compound	R_1	R_2	$-\Delta G$, kcal/mol
4.1	$HC\equiv C-$	(phenyl)	12.1 ± 0.1
4.2	$CH_3-C\equiv C-$	(phenyl)	13.9 ± 0.2
4.3	$H_2C=C(CH_3)-C\equiv C-$	(phenyl)	14.3 ± 0.1
4.4	$CF_3-C\equiv C-$	(phenyl)	14.2 ± 0.2
4.5	CF_3	(phenyl)	13.7 ± 0.3
4.6	CF_3	CF_3	11.7 ± 0.3
4.7	(phenyl)	(phenyl)	14.3 ± 0.2
4.8	(2-thiazolyl)	(phenyl)	14.1 ± 0.2
4.9	(methylcyclopentadienyl)	(phenyl)	13.9 ± 0.2
4.10	(phenyl–C≡C–)	(phenyl)	11.6 ± 0.2
4.11	(2-thienyl)	(2-thienyl)	14.6 ± 0.2
4.12	C_3H_7	(2-thienyl)	13.3 ± 0.1
4.13	iso - C_3H_7	(2-thienyl)	13.6 ± 0.1
4.14	iso - C_4H_9	(2-thienyl)	13.6 ± 0.2
4.15	n-C_6H_{13}	(2-thienyl)	12.8 ± 0.1
4.16	(cyclopentyl)	(2-thienyl)	14.1 ± 0.1
4.17	(cyclohexyl)	(2-thienyl)	13.5 ± 0.1

(*continued*)

TABLE 4.2 (Continued)
mAChR Receptors Binding Free Energy of Compounds with the Common Structure
Depicted in Figure 4.1 (Source: [12])

Compound	R$_1$	R$_2$	$-\Delta G$, kcal/mol
4.18	CH$_3$—[thienyl]	[thienyl]	14.6 ± 0.2
4.19	C$_2$H$_5$—[thienyl]	[thienyl]	13.9 ± 0.2
4.20	(CH$_3$)$_3$C—[thienyl]	[thienyl]	12.7 ± 0.1
4.21	C$_6$F$_5$	[thienyl]	12.9 ± 0.2
4.22	Br—[thienyl]	Br—[thienyl]	13.9 ± 0.1
4.23	[furyl]	[furyl]	13.2 ± 0.2

As a suitable reference compound in $\Delta\Delta G$ calculation, compound **4.1** (Table 4.2) was selected because of the small size of its substituting ethynyl group and the stable conformation. The ethynyl group's E$_{disp}$ is -1.0 kcal/mol.

All compounds with chiral atoms existed in the racemic form. To compare chiral compounds (R$_1$ $\neq$ R$_2$) with achiral ones (R$_1$ = R$_2$), it was necessary to use the ΔG value of an active isomer. The cholinolytic activity of (R)-isomers of amino glycolates is much higher than that of (S)-isomers [14, 15]; therefore, a difference in ΔG values between racemates and (R)-isomers is $\Delta\Delta G = 1.42 \log 2 \approx 0.4$ kcal/mol. The value of ΔG for an (R)-isomer of compound **4.1** is -12.5 kcal/mol. The contribution of a phenyl group to the free energy change can be estimated as -2.8 kcal/mol from comparing compounds **4.1** and **4.7** (Table 4.2). The substitution of two phenyl groups for thienyl groups (e.g., for compounds **4.7** and **4.11**) yields $\Delta\Delta G = -0.3$ kcal/mol. Hence, the contribution of thienyl groups can be estimated as -2.95 kcal/mol with an assumption that both equally contribute to the energy increment.

All chemical substituents presented in Table 4.3 can be subdivided into the three groups:

1. The first group contains those residues (group no. **2–9** as listed in Table 4.3) that have similar E$_{disp.}$ and $\Delta\Delta G$ values, with the difference being less than 0.5 kcal/mol. This group includes planar substituents except for cyclohexyl. It is quite probable that their position on the receptor fits the most to the model. These substituents are placed presumably on the receptor surface and interact with the receptor only from one side.

2. The second group contains substituents no. **10–16** (Table 4.3), whose $\Delta\Delta G$ is much greater than E$_{disp.}$ in magnitude. They can be incorporated closer to the protein backbone of the receptor's binding site; therefore, the actual binding strength is higher than the calculated value. Evidently, such an "incorporation effect" is impossible in the case of planar aromatic moieties belonging to the first group. The trifluoromethyl group in compounds **4.5** and **4.6** makes a much greater contribution than the predicted value. Perhaps, because of the high electronegativity of this group, the electron density distribution on neighboring atoms in the molecule is altered to facilitate stronger binding.

3. The third group comprises substituents with $\Delta\Delta G$ values that are significantly less in magnitude than E$_{disp}$ (no. **17–22** in Table 4.3). The difference in these parameters is related to steric barriers for binding because of the large size of these groups.

TABLE 4.3
Partial Contributions of Groups from Table 4.2 to Free Energy and Calculated Dispersion Interaction Energy. Reference Compound Depicted in Figure 4.1 (Table 4.2)

N	Group (for Nonsymmetrical Residues, Connecting Bond Is Shown)	$-\Delta\Delta G$ kcal/mol	Dispersion Energy $-E_{disp}$ kcal/mol	Radius of the Group r, Å	Molar Refraction of the Receptor Group, ml/mol	Number of Valence Electrons, Z	Equilibrium Distance, d_e Å
1	HC≡C—	-	1.00	1.70	8.68	10	3.70
2	(phenyl)	2.8	3.25	1.80	25.11	30	3.80
3	(thiophene)	2.95	2.93	1.82	24.25	26	3.82
4	(thiazole)	3.0	2.82	1.82	22.58	26	3.82
5	(furan)	2.25	2.27	1.80	18.21	26	3.80
6	(cyclopentadienyl)	2.8	2.63	1.80	21.09	26	3.80
7	(methylthiophene)	3.35	3.49	1.82	28.80	32	3.82
8	H₂C=C—C≡C—CH₃	3.2	3.20	1.70	21.90	26	3.70
9	(cyclohexane)	2.25	2.12	2.15	26.60	36	4.15

(continued)

TABLE 4.3 (Continued)
Partial Contributions of Groups from Table 4.2 to Free Energy and Calculated Dispersion Interaction Energy. Reference Compound Depicted in Figure 4.1 (Table 4.2)

N	Group (for Nonsymmetrical Residues, Connecting Bond Is Shown)	$-\Delta\Delta G$ kcal/mol	Dispersion Energy $-E_{disp}$ kcal/mol	Radius of the Group r, Å	Molar Refraction of the Receptor Group, ml/mol	Number of Valence Electrons, Z	Equilibrium Distance, d_e Å
10	C_3H_7-	2.05	1.30	2.0	15.15	20	4.00
11	iso - C_3H_7-	2.35	1.30	2.0	15.15	20	4.00
12	iso - C_4H_9-	2.35	1.80	2.0	19.65	26	4.00
13	(cyclopentyl)	2.85	2.10	2.0	22.10	30	4.00
14	$CH_3-C\equiv C-$	2.8	1.70	1.70	13.18	16	3.70
15	$CF_3-C\equiv C-$	3.1	1.81	1.80	13.57	34	3.80
16	CF_3-	2.6; 1.5	0.45	2.2	6.54	26	4.20
17	(phenyl $C\equiv C$)	0.5	4.41	1.80	32.14	38	3.80
18	C_6F_5-	1.65	3.95	1.80	25.76	60	3.80
19	(Br-thiophene)	2.6	3.74	1.88	31.79	32	3.88
20	$(CH_3)_3C$ (thiophene)	1.6	2.24	2.50	42.25	50	4.50
21	C_2H_5 (thiophene)	2.7	3.79	1.92	33.30	38	3.92
22	$CH_3(CH_2)_5-$	1.55	2.89	2.0	28.65	38	4.00

Among the considered antagonists, the (R)-isomer of compound **4.18** is the most active ligand ($\Delta G = -15$ kcal/mol). A possible reason for its prominent activity is that 5-methylthienyl (related to the first group of substituents) binds to the receptor without spatial obstacles, in contrast to its nearest homologs, namely, 5-ethylthienyl and 5-tert-butylthienyl, with both having the size greater than the optimal one. To the most extent, the steric hindrances appear during the binding of phenylethinyl, which has the largest size (of the regarded groups), in a bond direction along the main part of a molecule. As it follows from compared binding efficiencies for groups **17**, **20**, and **21** in Table 4.3, the group linear size of >6–7 Å is a prerequisite for steric hindrances that prevent an increase in the affinity to the receptor. Even small substituents (e.g., bromide atoms) to be introduced into both R_1 and R_2 groups would also cause steric barriers for ligand–receptor complex formation or intramolecular strains. Likely, n-hexyl (no. **22**) may adopt a folded (twist) configuration on the receptor surface with an appreciable entropy loss that decreases its partial contribution to the formation energy of free complex.

Thus, dispersion energy estimation can predict the affinity for mAChR in the considered series of ligands. However, the possibilities of exact predictions are limited by the size of the substituents, which must not exceed 6–7 Å. If the substituents belong to the second or third group, then the used model is valid only for an approximate estimation of the affinity of new compounds.

Noteworthy, a similarity between $\Delta\Delta G$ and $E_{disp.}$ is not so important, bearing in mind very approximate calculations. It is rather significant that the partial contributions of groups R_1 and R_2 to the free energy are changed depending on their van der Waals radii, molar refraction, and the number of valence electrons on the groups, with the size allowing the ligand to bind to the receptor without steric barriers. This dependence enables the judgment about whether a bound group is positioned on the receptor surface or immersed in the binding pocket to some depth and whether the steric barriers appear at binding.

4.4 ON APPLICABILITY OF METHODS FOR THE MEASUREMENT OF BINDING ENTHALPY AND ENTROPY

To gain deep insight into mechanisms involved in the interactions between ligands and receptors, it is useful to determine not only binding free energy but also binding enthalpy and entropy separately, although the measurements can be difficult. The combination of binding enthalpy and entropy ($\Delta H, -T\Delta S$) is commonly called "thermodynamic signature," as it is a suitable measure of a ligand binding profile used in drug discovery [16, 17]. Ligand binding to a protein is characterized by the sum of small contributions to entropy and enthalpy from various groups.

Enthalpy can be measured using calorimetric methods. Calorimetry-based measurements require purified receptors or cloned receptors, and they are not always available. Among the applied methods, isothermal titration calorimetry (ITC) is the most commonly one used [18, 19]. The method is based on directly measuring heat (enthalpy) related to the interaction of a protein and a ligand in a small sample reactor, in comparison with that of a reference reactor in an environment to ensure isothermal conditions. Increasing the ligand-to-protein molar ratio through ligand titration, the saturation is achieved. Nonlinear regression of the resulting sigmoidal curve is used to estimate enthalpy, K_d, and stoichiometry of the binding. ITC is widely used in the drug discovery process, where high throughput is needed. The applications of ITC to drug discovery are discussed in Chapter 3.

Enthalpy can be also obtained from temperature dependence of the equilibrium constant K_d using van't Hoff equation:

$$\frac{d\ln K_d}{d(1/T)} = -\frac{\Delta H}{R} \qquad (4.9)$$

Use of the known ΔH value enables the calculation of the entropy as

$$\Delta S = \frac{\Delta H - \Delta G}{RT} \tag{4.10}$$

The results of ΔH and ΔS measurements according to the van't Hoff equation for ligand-mAChR binding (Table 4.4) show less applicability of such calculations for ligand–receptor interactions.

TABLE 4.4
Thermodynamic Parameters of Ligand Binding to mAChR*

Compound	Optical Form	ΔG, kcal/ mol	ΔH, kcal/mol	$-T\Delta S$, kcal/mol	T range, °C	Refs
Atropine (Figure 4.2)	–	−13.5	−29.2	+16	30-37	[20]
	+	−10.8	−6.7	−4		
C-4.24	–	−13.6	−2.3	−11	30-37	[20]
C-4.25	–	−13.8	−9.4	−4	30-37	[20]
	+	−10.2	+2.8	−13		
C-4.26	–	−13.7	+6.0	−20	30-37	[20]
	+	−11.1	+5.4	−17		
C-4.27	–	−12.4	−3.1	−9	30-37	[22]
QNB (Figure 4.3)	Racemate	−13.6	+13.0	−26.6	29-37	[21]
QNB iodomethylate (Figure 4.4)	Race-mate	−10.9	−9.1	−1.8	29-37	[22]
Scopolamine (Figure 4.5)	–	−13.1	−17.1	+ 4.0	29-37	[24]
N-methyl-scopolamine (NMS) iodide (Figure 4.6)	–	−14.0	−36.4	+22.4	29-37	[24]

* The data were obtained from experiments on guinea pig small intestine, except for QNB where rat brain homogenate was used.

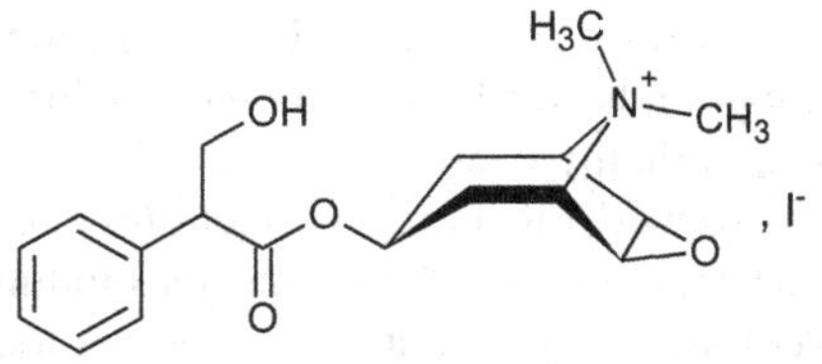

FIGURE 4.2 Structure of atropine iodomethylate.

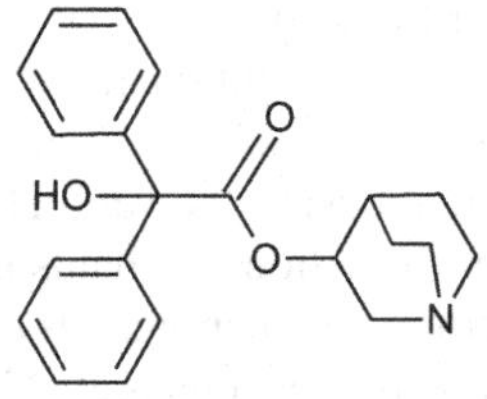

FIGURE 4.3 Structure of QNB.

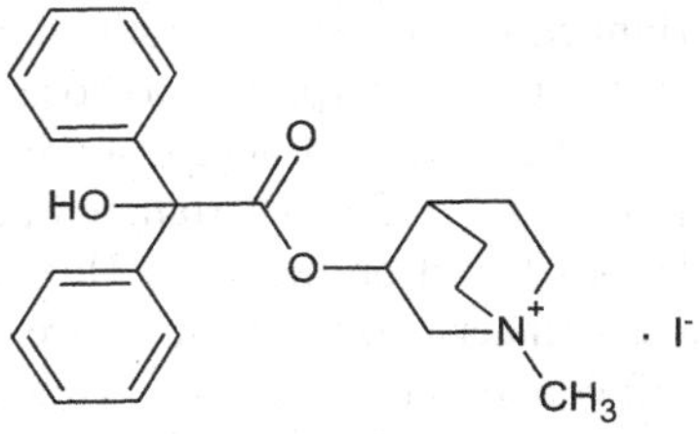

FIGURE 4.4 Structure of QNB iodomethylate.

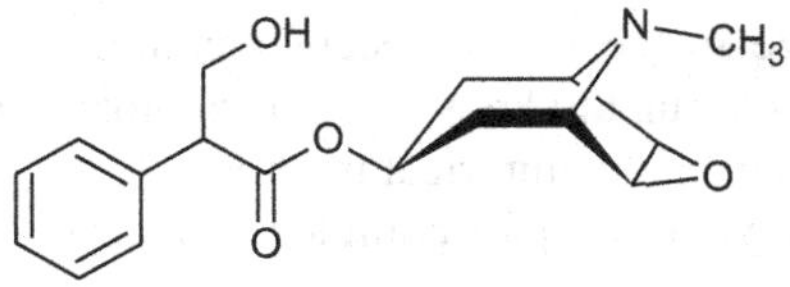

FIGURE 4.5 Structure of scopolamine.

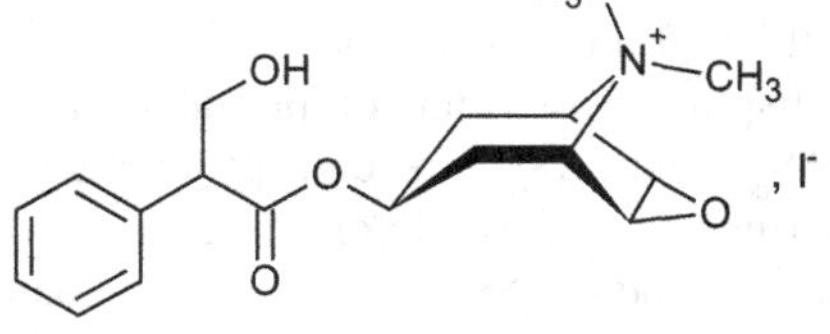

FIGURE 4.6 Structure of N-methyl-scopolamine (NMS) iodide.

No noticeable pattern was observed for changes in ΔH and $T\Delta S$: complex formation is profitable in terms of entropy for some compounds, while for the others, which have a similar structure, this process is enthalpy driven. Thus, enthalpy change ($\Delta H = -17.1$ kcal/mol) is the major factor for scopolamine (Figure 4.6) binding to mAChR. For 3-quinuclidinyl benzilate (Figure 4.4) as another selective mAChR ligand, the ΔH is $= +13$ kcal/mol, and its binding to the receptor occurs as a result of an entropy gain only. As for (+)- and (−)-stereoisomers of iodoethylated diethylaminoethyl ester of phenylcyclohexylglycolic acid C-**4.26**, the binding to the receptor is favored by the entropy change. Again, an increased entropy is the major contributor to complex formation between the (+)-isomer of the trimethylammonium analog C-**4.25** and mAChR, but, by contrast, the change in enthalpy is conditional for its (−)-isomer binding.

The applicability of the van't Hoff equation for interactions with proteins was discussed in detail by Blumenfeld [23], who called attention to discrepancies between the true ΔH values, measured using the calorimetric method, and the values obtained using the van't Hoff law. To use the van't Hoff equation correctly, it is assumed that the complex formation mechanism remains the same in the chosen temperature range, but this is not true for binding between ligands and proteins as noted by Blumenfeld, as proteins may undergo considerable conformational changes. The author concluded that the difference in experimentally assessed and steady-state calculated heat effects reflects the appreciable conformational rearrangements in protein molecules even during slight temperature changes.

Thus, the van't Hoff analysis method describes the dependence of transitions in the protein conformation (rather than the equilibrium constant) on temperature change. Noteworthy, the results to be obtained for receptors are more difficult to interpret than those to be obtained for enzymes. The data listed in Table 4.7 possibly suggest the effect of temperature on an entire membrane–receptor complex. A linear relationship of $\ln K$ versus $1/T$ in a certain temperature range is often considered as an argument for correct calculations based on the van't Hoff equation. However, as noted by Blumenfeld, conformational changes in macromolecular structure can proceed in a gradual manner. It is quite possible that the van't Hoff equation is applicable to the thermodynamic analysis of some ligand–receptor systems, but these cases should be carefully validated.

4.5 ENTROPY TERMS

In experiments, the total change in the ligand–receptor complex formation entropy can also be determined, although it is also important to know the contributors to the total change ΔS, which can be estimated using semi-quantitative and empirical methods.

The entropy change during ligand–receptor complex formation can be represented as a sum of the following terms [24]:

$$\Delta S = \Delta S_{rot} + \Delta S_{tr} + \Delta S_{conf} + \Delta S_{osc} + \Delta S_w \qquad (4.11).$$

where ΔS_{rot} and ΔS_{tr} are the rotational and translational entropy changes, respectively; ΔS_w is the entropy change during water displacement (total or partial removal of hydrate shells from the receptor's binding site); and ΔS_{conf} and ΔS_{osc} are entropy changes pertaining to the ligand and receptor conformational transformation and oscillatory degrees of freedom, respectively. For most ligand–receptor complexes, $-T\Delta S_{rot}$ and $-T\Delta S_{tr}$ values are approximately 2 kcal/mol and 5 kcal/mol, respectively, although these values are negative because the number of translational and rotational degrees of freedom is declining [25]. $\Delta S_{appr} = \Delta S_{tr} + \Delta S_{rot}$ is a determinant of the entropy for approaching molecules. The $-T\Delta S_{appr}$ value for an interaction between an amino acid residue and an enzyme at room temperature is ~2–4 kcal/mol [26].

Hydrogen bonding between a ligand and its receptor can impede intramolecular oscillations and produce new low-frequency oscillations. As a rule, ΔS_{osc} is supposed to be positive, that is, the total oscillation entropy increases because new oscillations appear in a weakly bound ligand–receptor complex. Numerically, this value is dependent on oscillations, having a frequency of not more than 500 cm^{-1}. $T\Delta S_{osc}$ is = 0.5 kcal/mol for each formed hydrogen bond [27].

The major contributor to the total entropy change may be the ΔS_{conf} value pertaining to the conformational entropy that depends on the number of variable torsional angles in a ligand molecule frozen in the complex. An estimated entropy loss resulting from attachment to the receptor of a carbon chain, having a torsion angle C^{α} - C^{β} - C^{γ} - C^{δ}, is approximately 1.7 kcal/mol (for torsion angles of free rotation) [27] or 0.7 kcal/mol [24]. This estimation is conditional upon the number of conformations (energy minima) corresponding to different values of the torsion angle. It is quite probable that contributions due to fixation of independent torsion angles are additive.

The release of water during ligand binding to the receptor is accompanied by an increase in entropy that is difficult to estimate exactly in view of the unknown extent of environment changes in the excluded water molecules. Such contribution to changes in the system's entropy is attributed to differences in the motion of water molecules in the hydrate shell and solution. A value of ΔS_w can be estimated from comparison of the molecular surface available for hydration of the free or complexed ligand provided if its structure is known. Often, the estimated ΔS_w value for water displacement can be written as the sum of contributions from single groups of a ligand molecule and its receptor depending on their polarity. The most polar side chains in amino acids make the highest contributions to the increase in system entropy. The maximum value $T\Delta S_w$ of 2.1 kcal/mol at 27 °C for asparagine is attainable if more than 60% of the ligand surface is immersed into the receptor binding pocket [25, 28]. Nonpolar groups contribute only little to the ΔS_w value. As estimated in [29, 30], a transition of both the ligand and its receptor from the free state to the bound state is characterized in the case of complete water displacement from the contact site by $T\Delta S_w$ values of no more than 5–6 kcal/mol for an ionic group (e.g. trimethylammonium) and up to 2–3 kcal/mol for a dipole group (e.g., carbonyl). Probably, these are overestimations, as the entropy of the association of the displaced water molecules, ΔS_{w-w}, having a negative value, should also be considered.

To calculate ΔS, the contributions of the receptor groups should be considered, but this is very difficult, particularly, because of many possible conformational states of the protein chain when compared with ligands.

Generally, the entropy change during ligand–receptor binding is contributed by the following factors:

1) The rigidity of a formed ligand–receptor complex resulting from the loss of translational degrees of freedom in a ligand molecule and the loss or restriction of rotational motions (a negative ΔS value).
2) Changes in the receptor and ligand structure resulting from conformational rearrangements when ΔS can be positive or negative.
3) The release of previously bound water molecules with entropy gain [9c].

4.6 SIMULATION METHODS FOR BINDING FREE ENERGY CALCULATION

Molecular dynamics (MD) approaches are widely used in the simulation of gases, liquids, nanoparticles, electrolytes, and supramolecular surfaces. Thermodynamics of the formation of the ligand–receptor complexes can also be investigated *in silico* using computer simulations. Methods of MD for the calculation of ΔG values and other quantities for ligand–receptor complexes have been rapidly developed in recent two decades [31-34]. Because of the significantly increased computer power and the wide availability of graphics processing unit for massive parallel computations, the commercial and open-source software packages are routinely used for microsecond-scale molecular simulations of protein–ligand binding. Generally, MD is based on the atomistic representation of a system by solving Newtonian equation of motion for each atom at each step of the system trajectory in the configurational phase space. Despite the huge dimension of phase space (~3N, where N is the number of atoms in the system), it can be reduced by division into cells with a ~100 Å linear scale and applying periodical boundary conditions. This, in turn, produces some artificial boundary effects, but they can be processed using standard techniques. Therefore, the obtained trajectory is a time series of snapshots with atomic positions, velocities, and forces.

MD assumes that, through long enough simulation, we can obtain a full series of microscopic states that gives reliable values of physical observables of the macroscopic system (heat of evaporation, radial distribution function, and density). Many factors influence the necessary length and accuracy of a simulation, among which are applied integrator in equation of motion, sampling procedures, selected force field correctness, amount of polarization effects, and algorithms for iterations.

Calculation of atomic forces requires knowledge about the interaction potential between atoms, and it is, in most cases, encoded in bonded potentials for atoms in molecules (bonds, angles, and torsions) and atomic pair potentials for nonbonded interactions (dispersion and electrostatics) within force fields—a set of parameters and potential functions. Force field parametrization might be time-consuming and needs extensive attention, but once done, it can be applied widely. Nowadays, both experimental data and *ab initio* quantum chemical (QC) data are used for parametrization, but there are force fields entirely relying on pure QC calculations [35-38]. A popular way is to apply machine learning methods for the construction of empirical potentials and corrections of force field errors because of nonadditivity of interactions and several body effects [39-41].

To calculate the $\Delta\Delta G$ value for two ligands or two residues in a target protein with the same ligand or wild-type (apo-protein) ligand, the method of the artificial alchemical thermodynamic cycle is applied, where a ligand (or protein residue) alchemically transforms (mutates) to its modification. It can be represented as the following thermodynamic cycle, taking unbound ligand A in water as the initial point:

In the scheme, two vertical paths are equivalent to the horizontal ones, so

$$\Delta\Delta G_{A-B} = \Delta G^B_{\text{Water-Protein}} - \Delta G^A_{\text{Water-Protein}} = \Delta G^{A-B}_{\text{Protein}} - \Delta G^{A-B}_{\text{Water}}$$

To calculate the latter difference, we need to simulate mutation only in water and protein. Technically, mutation is represented by the dimensionless parameter λ in interaction potential:

$$V(r,\lambda) = f(\lambda)V(r), 0 < \lambda < 1$$

$$V(r,0) = V(r)^A, \ V(r,1) = V(r)^B$$

$$\lambda = \left(0, \lambda_1, \lambda_2, \cdots \lambda_{n-1}, 1\right)$$

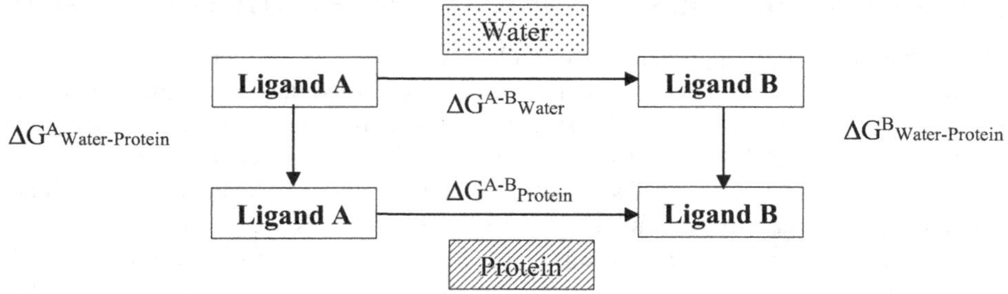

FIGURE 4.7   Ligand transformation cycle to obtain $\Delta\Delta G$.

Hence, instead of direct binding simulation, we perform the artificial ligand mutation from A to B in water and protein. The mutation simulation is statistically more feasible than direct binding simulation. This simulation is performed for each λ value from the specified discrete set of values from 0 to 1. The more the λ values are used, the better the statistics, and commonly, no less than 10 points are used.

MD trajectory length is measured in total length of the time series (product of time step length on number of steps), and acceptable values are at least 1 ns. The convergence of the trajectory should be checked under various conditions [42]. Once trajectories for each λ value are obtained, ΔG can be calculated using standard methods of statistical physics, among which are thermodynamics integration and Bennett acceptance ratio. There are easily accessible open-source software packages where the methods are implemented.

A similar alchemical method is used to compute ΔG absolute binding free energy [43]; albeit, it is more sophisticated and is not considered here.

The accuracy of MD methods depends mainly on force field accuracy, quality of force field optimization, and configurational sampling. Recent benchmarks have revealed an average unsigned error of ~1 kcal/mol for $\Delta \Delta G$ [44, 45], which might be enough for high-throughput virtual screening of lead compounds. In the best cases, the obtained mean average error for $\Delta \Delta G$ in a series of ligands can be less than 1 kcal/mol. It was concluded in [46] that for MD relative binding free energy methods, an accuracy of 0.5 kcal/mol is still far from reaching.

Free energy binding calculations are still at the beginning of applying on actual protein–ligand biomolecular systems, but undoubtedly, great progress will be achieved in this decade in terms of the speed and precision of calculations while improving the physical accuracy of force fields, simulation methods, and configuration sampling.

NOTES

1 By Albert [42], "pharmacodynamics agents" are substances acting on the organism's native cells; on the contrary, chemotherapeutics act on pathogens.
2 Atropine, scopolamine, adiphenine, 3-quinuclidinyl benzilate, and other substances have low selectivity to mAChR subtypes [43]

REFERENCES

[1] D. Metzler, Biochemistry, New York: Academic Press, 2003, p. 244.
[2] M. Watson, W. Roeske and H. Yamamura, *Life Sci.*, vol. 31, p. 2019, 1982.
[3] E. Hulme, N. Birdsall, A. Burgen and P. Menta, *Mol. Pharmacol.*, vol. 14, p. 737, 1978.
[4] E. Hulme, N. Birdsall and N. Buckley, *Ann. Rev. Pharmacol. Toxicol.*, vol. 30, p. 633, 1990.
[5] A. Michel and R. Whiting, *Brit. J. Pharmacol.*, vol. 92, p. 755, 1987.
[6] M. Mashkovskii, Drugs (in Russian), vol. 1, Moscow: Novaya Volna, 2021.
[7] L. Pauling and D. Pressman, *J. Am. Chem. Soc.*, vol. 67, p. 1003, 1945.
[8] J. Webb, Enzyme and metabolic inhibitors, New York: Academic Press, 1963, (a) p. 264, (b) p. 272, (c) p. 261.
[9] A. Nisonoff and D. Pressman, *J. Am. Chem. Soc.*, vol. 79, p. 1616, 1957.
[10] D. Pressman and M. Siegel, *J. Am. Chem. Soc.*, vol. 75, p. 686, 1953.
[11] I. Kloog and M. Sokolovsky, *Biochem. Biophys. Res. Commun.*, vol. 81, p. 710, 1978.
[12] F. Dukhovich, E. Gorbatova and V. Kurochkin, *Pharm. Chem. J.*, vol. 35, p. 260, 2001.
[13] F. Dukhovich, E. Gorbatova, M. Darkhovskii and V. Kurochkin, Molecular Recognition: Pharmacological Aspects, New York: Nova Science Publishers, 2004.
[14] J. Wijngaarden, W. Soudyn and C. Eycken, *Life Sci.*, vol. 15, p. 1289, 1970.
[15] R. Brimblecombe, D. Green, T. Inch and P. Thompson, *J. Pharm. Pharmacol.*, vol. 23, p. 745, 1971.
[16] E. Freire, *Chem. Biol. Drug Des.*, vol. 74, no. 5, pp. 468–472, 2009.
[17] J. Chaires, *Annu. Rev. Biophys.*, vol. 37, no. 1, pp. 135–151, 2008.

[18] I. Jelesarov and H. Bosshard, *J. Mol. Recognition*, vol. 12, pp. 3–18, 1999.

[19] G. Holdgate and W. Ward, *Drug Discovery Today*, vol. 10, no. 22, pp. 1543–1550, 2005.

[20] R. Barlow and K. Burston, *Brit. J. Pharmacol.*, vol. 66, p. 581, 1976.

[21] R. Aronstam, L. Abood and J. Baumgold, *Biochem. Pharmacol.*, vol. 26, p. 1689, 1977.

[22] R. Barlow, K. Berry, P. Glenton, N. Nikolaou and K. Soh, *Br. J. Pharmacol.*, vol. 58, p. 613, 1976.

[23] L. Blumenfeld, Problems of Biological Physics, Berlin: Springer, 1981, p. 120.

[24] M.A. Murcko, *J. Med. Chem.*, vol. 38, p. 4953, 1995.

[25] J. Novotny, R. Bruccoleri and F. Saul, *Biochemistry*, vol. 28, p. 4735, 1989.

[26] M. Bender, F. Kezdy and C. Gunter, *J. Am. Chem. Soc.*, vol. 86, p. 3714, 1964.

[27] S. Vajda, Z. Weng, R. Rosenfeld and C. De Lisi, *Biochemistry*, vol. 33, p. 13977, 1994.

[28] J. Novotny, R. Bruccoleri, M. Davis and K. Sharp, *J. Mol. Biol.*, vol. 268, p. 401, 1997.

[29] J. Dunitz, *Science*, vol. 264, p. 670, 1994.

[30] W. Bryan, *Science*, vol. 266, p. 1726, 1994.

[31] A. Wade, A. Bhati, S. Wan and P. Coveney, *J. Chem. Theory Comput.*, vol. 18, no. 6, pp. 3972–3987, 2022.

[32] Z. Cournia, B. Allen and W. Sherman, *J. Chem. Inf. Model.*, vol. 57, no. 12, pp. 2911–2937, 2017.

[33] J. Chodera, D. Mobley, M. Shirts, R. Dixon, K. Branson and V. Pamde, *Curr. Opin. Struct. Biol.*, vol. 21, p. 150, 2011.

[34] J. Mortier, C. Rakers, M. Bermudez, M. Murgueitio, S. Riniker and G. Wolber, *Drug Discov. Today*, vol. 20, p. 686, 2015.

[35] A.G. Donchev, N.G. Galkin, A.A. Illarionov, et al., *J. Comput. Chem.*, vol. 29, p. 1242, 2008.

[36] C.S. Ewig, R. Berry, U. Dinur, et al., *J. Comput. Chem.*, vol. 22, p. 1782, 2001.

[37] S.A. Grimme, *J. Chem. Theory Comput.*, vol. 10, p. 4497, 2014.

[38] L. Pereyaslavets, G. Kamath, O. Butin, et al., *Nat. Commun.*, vol. 13, p. 414, 2022.

[39] J. Behler, *J. Chem. Phys.*, vol. 145, p. 170901, 2016.

[40] C. Chen and S. Ong, *Nat. Comp. Sci.*, vol. 2, p. 718, 2022.

[41] J. Smith, O. Isayev and A. Roitberg, *Chem. Sci.*, vol. 8, p. 3192, 2017.

[42] P. Klimovich, M. Shirts and D. Mobley, *J. Comput. Aided. Mol. Des.*, vol. 29, no. 397, 2015.

[43] M. Aldeghi, A. Heifetz, M. Bodkin, S. Knapp and P. Biggin, *Chem. Sci.*, vol. 7, p. 207, 2016.

[44] D. Mobley and M. Gilson, *Annu. Rev. Biophys.*, vol. 46, p. 531, 2017.

[45] V. Gapsys, D. Hahn, G. Tresadern, D. Mobley and B. de Groot, *J. Chem. Inf. Model.*, vol. 62, p. 1172, 2022.

[46] D. Mondal, J. Florian and A. Warshel, *J. Phys. Chem. B,* vol. 123, p. 8910, 2019.

5 Effect of Conformational Entropy on Affinity of Specific Ligands

5.1 METHODS FOR THE CALCULATION OF CONFORMER NUMBER AND CONFORMATIONAL ENTROPY

To calculate conformational entropy, the number of conformers should be determined. We used the TINKER molecular modeling program [2] to calculate the conformation set. The procedure for the conformational analysis of all molecules under study is based on searching for the potential energy minima in main vibrational modes that correspond to changes in certain torsion angles. To calculate the potential energy, MM3 parametrization [3] is used. All the potential energy components used in this parameterization are considered. Calculations were performed for the cationic form in which the molecules bind to the receptor.

To search for the conformers, we scanned the torsion angles along the main chain in the ester fragment of glycolates and acetates, which are substituents at the C-alpha atom of the acidic part and alkyl groups containing the nitrogen atom. In these settings, we generated conformations at an energy threshold value of 10 kcal/mol.

The ΔS_{conf} value was derived based on the assumption that a ligand binds to its receptor at one complementary conformation. Then, the conformational entropy change in the formation of the ligand–receptor complex is expressed through Boltzmann distribution of all rotamers [4]:

$$\Delta S_{conf} = -R \sum_{i}^{N_{conf}} p_i \ln p_i,$$

where R is the gas constant, N_{conf} is the number of conformers of an unbound molecule, i is the ordinal number of a conformer, and p_i is the probability of the i-th conformer in an equilibrium system:

$$p_i = \frac{e^{-\beta E_i}}{\sum e^{-\beta E_i}},$$

where E_i is the potential energy of the i-th conformer, and $\beta = 1/kT$ (k is the Boltzmann constant). The p_i value is calculated at 310.5 K (37.5 °C) where the experimental K_d (ΔG) values are determined.

Noteworthy, ΔS_{conf} is not proportional to the number of conformations (N_{conf}), as the distribution of the p_i values is far from uniform. Conformational changes are not localized in one fragment wherein these changes occurred but rather propagate to the whole molecule. Therefore, modifying one fragment can usually cause an increase or decrease in N_{conf} pertaining to the whole molecule.

5.2 SPECIFIC LIGANDS OF M-ACETYLCHOLINE RECEPTORS

As the substances (Figure 5.1) were available as racemates, the ΔG values presented in Table 5.1 represent the most potent (R)- or (R,R)-form for better comparison with nonchiral substances. Affinities of (R)-isomers of the chiral amino esters of the substituted glycolic and acetic acid compounds greatly dominate those of (S)-isomers [5]. Owing to possible differences in the binding mode (see detailed discussion in Chapter 6), glycolates and acetates are considered separately.

Data presented in Table 5.1 enable the calculation of entropy change upon varying conformational flexibility of the substances. Conformational flexibility varies through two ways:

- Molecular structural framework modification during the substitution of more conformationally stable N-methyl-4-piperidyl and 3-quinuclidinyl groups for N-alkyl groups
- Substitution of conformationally rigid phenyl and thienyl groups for cyclohexyl.

The data presented in Table 5.1 suggest the possibility to compare calculated conformational entropy $\Delta(T\Delta S_{conf})$ and free energy ΔG changes with varied conformational rigidity of a

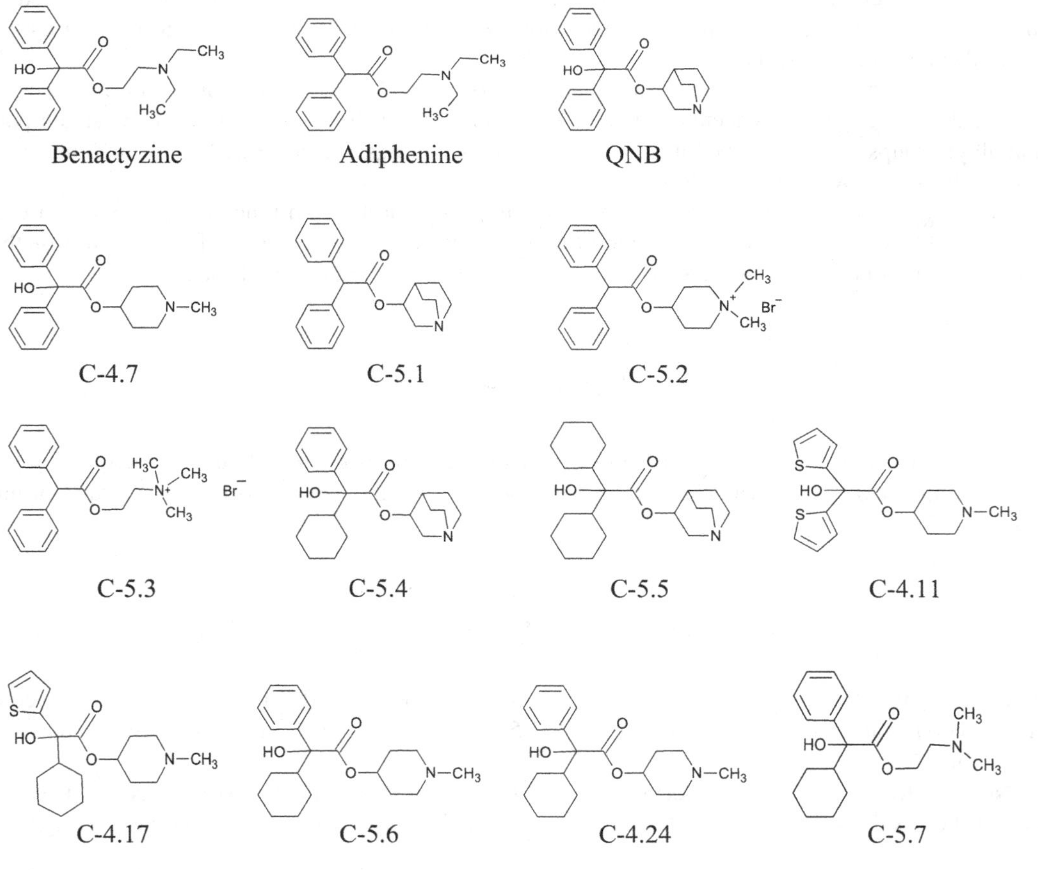

FIGURE 5.1   Specific ligands of mAChR.

TABLE 5.1
Number of Conformations and Conformational Entropy Changes on Structural Modification of Substances[1]

Substance	N_{conf}	$-T\Delta S_{conf}$, kcal/mol	$\Delta(T\Delta S_{conf})$, kcal/mol	$-\Delta G$, kcal/mol	$-\Delta\Delta G$, kcal/mol	$\dfrac{\Delta\Delta G}{\Delta(T\Delta S_{conf})}$
QNB	12	0.3	1.9	14.4	2.3	1.2
C-4.7	12	0.4	1.8	14.3	2.2	1.2
Benactyzine	451	2.2	-	12.1		
C-5.1	5	0.3	2.0	14.2	3.6	1.8
Adiphenine	415	2.3	-	10.6		
C-5.2	53	0.7	1.2	12.8	2.4	2.0

(*continued*)

TABLE 5.1 (Continued)
Number of Conformations and Conformational Entropy Changes on Structural Modification of Substances[1]

Substance	N_{conf}	$-T\Delta S_{conf}$, kcal/mol	$\Delta(T\Delta S_{conf})$, kcal/mol	$-\Delta G$, kcal/mol	$-\Delta\Delta G$, kcal/mol	$\dfrac{\Delta\Delta G}{\Delta(T\Delta S_{conf})}$
C-5.3	75	1.9	-	10.4		
QNB	12	0.3	1.2	14.4	2.1	1.2
C-5.4	327	1.5	0.5	13.5	1.2	2.4
C-5.5	2329	2.0	-	12.3		
C-4.11	84	0.9	0.7	14.6	1.1	1.6
C-4.17	610	1.6	-	13.5		

TABLE 5.1 (Continued)
Number of Conformations and Conformational Entropy Changes on Structural Modification of Substances[1]

Substance	N_{conf}	$-T\Delta S_{conf}$, kcal/mol	$\Delta(T\Delta S_{conf})$, kcal/mol	$-\Delta G$, kcal/mol	$-\Delta\Delta G$, kcal/mol	$\dfrac{\Delta\Delta G}{\Delta(T\Delta S_{conf})}$
C-4.7	8	0.4	0.3	14.3	0.7	2.3
C-5.6	103	0.7	-	13.6		
C-4.24	103	0.7	1.0	13.6	1.2	1.2
C-5.7	808	1.7	-	12.4		

compound. The following series of substances from Figure 5.1 were compared: (QNB, C-**4.7**, benactyzine), (C-**5.1**, adiphenine), (C-**5.2**, C-**5.3**), (QNB, C-**5.4**, C-**5.5**), (C-**4.11**, C-**4.17**), (C-**4.7**, C-**5.6**), and (C-**4.24**, C-**5.7**). An increase in the ligand's structural rigidity is accompanied by declining conformational entropy, and therefore, the selected calculation model adequately "reacts" to modifications in the structure of the substances. In the first direction of the proposed flexibility changes, the transition from the relatively labile benactyzine to more rigid glycolates C-**4.7**, or QNB reduces the number of conformations N_{conf} from 451 to 12, and the value of $T\Delta S_{conf}$ from 2.2 kcal/mol to 0.4 and 0.3 kcal/mol, respectively. Similar structural modifications in ligands from acetic acid derivatives such as C-**5.1** and adiphenine are accompanied by the reduction of N_{conf} from 415 to 5 and the value of $T\Delta S_{conf}$ from 2.3 kcal/mol to 0.3 kcal/mol. A similar pattern of N_{conf} and $T\Delta S_{conf}$ changes occurs for substances with the quaternary nitrogen atom (C-**5.2** and C-**5.3**).

The second scenario with the substitution of a phenyl (thienyl) group for cyclohexyl is followed by increasing N_{conf} values from 12 to 327 and $T\Delta S_{conf}$ values from 0.3 kcal/mol to 1.5 kcal/mol

(QNB and compound C-**5.4**). With the replacement of both phenyl rings for cyclohexyl, the N_{conf} value sharply increases to 2329 and the $T\Delta S_{conf}$ value increases to 2.0 kcal/mol (QNB and compound C-**5.5**). Analogous changes are obtained in pairs (C-**4.11** and C-**4.17**) and (C-**4.7** and C-**5.6**). Thus, conformational entropy might be used to forecast substances' affinity to the receptor.

The most interesting result followed from Table 5.1 is that the value of $\Delta\Delta G$ is nearly twice the value of $\Delta\left(T\Delta S_{conf}\right)$; the average ratio $\Delta\Delta G / \Delta\left(T\Delta S_{conf}\right)$ is 1.7, and the standard deviation is 0.49.

From this relationship, the following conclusions were drawn:

• Loss of conformation entropy is distributed nearly equally between mAChR and the antagonists studied
• Conformational changes in mAChR during complex formation are localized in the binding site of the receptor and do not extend beyond it.

Almost the same degree of change is observed in the conformation entropy of both antagonists and receptors because their binding is based on the principle of complementarity. Differences in conformation entropy can be expected when acetylcholine and other agonists act on the receptor.

It is worth mentioning that only benactyzine and adiphenine (Figure 5.1) are used as pharmaceuticals [6]. The remaining substances have too strong affinity to mAChR to be used as drugs (see discussion in Chapter 6).

5.3   CONFORMATIONAL ENTROPY: ANTIDEPRESSANTS

The dependence of conformational entropy on structural rigidity can be investigated for other classes of pharmacological agents. In Table 5.2, the results of N_{conf} and $T\Delta S_{conf}$ calculations are used for some antidepressants (Figure 5.2), inhibitors of neurotransmitters' reverse uptake, and antagonists of D_2-dopamine receptors. Following the same principle as that we use for screening ligands of the mAChR receptor, we choose substances in a way such that changes in enthalpy and entropy factors should be minimized but allowing conformational rigidity to vary upon changes in the structure of the substances.

Substitutions of a single bond for a double bond and of a methylene chain for the conformationally stable N-methyl-4-piperidyl group contribute to the conformational rigidity of the substances (Figure 5.2). These structural changes produce the expected decrease in N_{conf} and $T\Delta S_{conf}$ values. Loss of conformational entropy is distributed nearly equally between the receptor and the antagonists (according to the $\Delta\Delta G / \Delta\left(T\Delta S_{conf}\right)$ ratio), in the same way as that in the interaction of mAChR with specific ligands. In this case, conformational changes in the receptor during complex formation are also localized in the binding site of the receptor.

TABLE 5.2
Conformational Entropy Changes in Some Antidepressants (Figure 5.3), Inhibitors of Neurotransmitters' Reverse Uptake Upon Binding to D_2-Dopamine Receptors

Substance	N_{conf}	$T\Delta S_{conf}$, kcal/mol	K_d, nM	$-\Delta G$, kcal/mol	$\dfrac{\Delta\Delta G}{\Delta\left(T\Delta S_{conf}\right)}$
Pizotifen	5	0.3	6.5 [7]	11.1	2.5
Cyproheptadine	11	0.5	31 [7]	10.2	2.2
Amitriptyline	23	0.7	210 [8]	8.9	1.4
Imipramine	85	1.6	2000 [9]	8.7	

5.4 CONFORMATIONAL ENTROPY AND ENDOGENOUS OPIATES

Conformational entropy may be considered as a universal factor in the preservation of the receptor's functional activity.

Values of the affinity and entropy changes during binding between morphine and met-enkephalin with μ-opioid receptors mediating the analgesic effect are summarized in Table 5.3. Met-enkephalin is a pentapeptide (*Tyr-Gly-Gly-Phe-Met*), acting as an endogenous μ-receptor agonist.

Meth-enkephalin, which had twofold higher molecular weight than morphine, slightly differs in affinity toward μ-receptors. Like morphine and morphine-like opiates, met-enkephalin is competitively displaced from μ-receptors by [³H]-naloxone, which proves the common binding site for these substances at the receptor. Unlike the conformationally rigid morphine molecule, the met-enkephalin molecule is conformationally labile; therefore, the N_{conf} and $T\Delta S_{conf}$ values are increased, consequently decreasing affinity. As pointed out before, those substances with $K_d < 0.1$ nM ($\Delta G < -14$ kcal/mol) irreversibly inactivate the receptor; hence, these drugs become poisonous. As shown from the data presented in Table 5.3, enkephalins would become toxic and poisonous without the influence of conformational entropy.

Endogenous ligands of opioid receptors are not limited to enkephalins. Endorphins are larger sizeable molecules. Thus, α- and β-endorphins, comprising 16 and 31 amino acid residues, are competitively displaced from μ-receptors by low-molecular-weight opiates and have relatively comparable affinities for these receptors, that is, 11 nM [10] and 3.2 nM [12], respectively.

In large molecules, even if not all conformations are frozen during binding, conformational entropy plays an important and likely major role in the preservation of the receptor's functional activity.

Smaller affinities of relatively large molecules are reasonably explained by steric hindrances, which may be undoubtedly true. However, the described relationship between $\Delta(T\Delta S_{conf})$ and $\Delta\Delta G$ indicates that entropy-based control of affinity is a major factor for conferring protection to neuroreceptors from irreversible inactivation when binding with large ligand molecules such as enkephalins, endorphins, hormones, and peptide drugs.

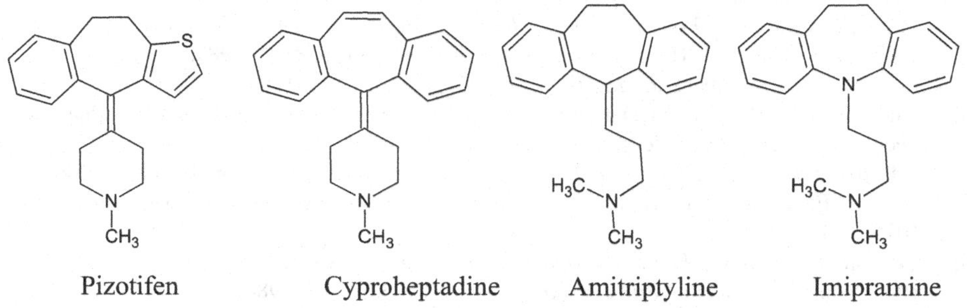

Pizotifen   Cyproheptadine   Amitriptyline   Imipramine

FIGURE 5.2 Structure of some of the common antidepressants.

TABLE 5.3
$T\Delta S_{conf}$ and ΔG values for Met-Enkephalin and Morphine Binding to Rat Brain μ-Opiate Receptors

Ligand	Molecular Weight	N_{conf}	$T\Delta S_{conf}$, kcal/mol	K_d, nM	$-\Delta G$, kcal/mol
Met-enkephalin	574	13329	3.84	3.0 [10]	11.6
Morphine	286	5	0.57	1.5 [11]	12.0

5.5 PROTEIN CONFORMATIONAL ENTROPY

The effect of receptor conformational entropy changes on affinity is equally important, but it cannot be as easily calculated as a ligand conformational entropy. In a previous study [13], theoretical estimation of conformational entropy loss due to protein folding was given for the main chain of amino acid residues. Each residue contributes to ~1 kcal/mol conformational entropy loss in energy units at 300 K, emerging from restricted rotations. Analysis of entropy changes is often reduced to solvation entropy estimations simply because no reliable methods are available to assess conformational entropy changes in the protein. On the contrary, Wand and coauthors ([14] and references therein) proposed and calibrated an experimental method to determine changes in protein conformational entropy, based on NMR relaxation measurements of dynamics of the protein methyl groups, called "entropy meter." Detection of fast internal side-chain motion helped the authors evaluate conformational entropy changes in 28 protein–ligand complexes with a broad range of binding affinities (from 0.1 mM to 0.1 nM). The protein conformational entropy changes range from +18 to −10 kcal/mol in different ligand–protein complexes; therefore, the influence of ΔS_{conf} on total affinity is important. The same method was applied later to provide insights into residual conformational entropy of the transmembrane proteins pSRII and OmpW [15] and to critically assess the predicted influence of water on protein conformational entropy [16].

Together with conformational protein dynamics extracted from X-ray crystallography datasets [17], the "entropy meter" can help understand the effects of protein conformational entropy on ligand binding. The role of conformational entropy in protein–ligand binding warrants further investigations.

NOTE

1   K_d (ΔG) data are taken from [1], except for compounds C-**5.3**, C-**5.4**: their K_d (ΔG) data are taken from [13].

REFERENCES

[1] C. Chang, W. Chen and M. Gilson, *Proc. Natl. Acad. Sci. USA*, vol. 104, p. 1534, 2007.

[2] R. Pappu, R. Hart and J. Ponder, *J. Phys. Chem. B.*, vol. 102, no. 48, p. 9725, 1998.

[3] N. Allinger, Y. Yuh and J.-H. Lii, *J. Am. Chem. Soc.*, vol. 111, p. 8551, 1989.

[4] S. Pickett and M. Sternberg, *J. Mol. Biol.*, vol. 231, no. 3, p. 825, 1993.

[5] F. Dukhovich, E. Gorbatova, M. Darkhovskii and V. Kurochkin, Molecular Recognition: Pharmacological Aspects, New York: Nova Science Publishers, 2004.

[6] M. Mashkovskii, Medical Compounds, vol. 1, Moscow, Novaya: Volna (in Russian), 2000, p. 140.

[7] J. Leysen, F. Awouters, L. Kennis, P. Laduron, J. Vandenberk and P. Janssen, *Life Sci.*, vol. 28, p. 1015, 1981.

[8] H. Hall and S.-O. Ogren, *Eur. J. Pharmacol.*, vol. 70, p. 393, 1981.

[9] E. Richelson and A. Nelson, *Eur. J. Pharmacol.*, vol. 103, p. 197, 1984.

[10] S. Childers, I. Creese, A. Snowman and S. Synder, *Eur. J. Pharmacol.*, vol. 55, no. 1, p. 11, 1979.

[11] S. Zaytsev, K. Yarygin and S. Varfolomeev, Drug addiction. Neuropeptide morphine receptors (in Russian), Moscow: Moscow University, 1993, p. 88.

[12] I. Goldberg, G. Rossi, S. Letchworth, et al., *J. Pharmacol. Exp. Ther.*, vol. 286, no. 2, p. 1007, 1998.

[13] D. Pal and P. Chakrabarti, *Proteins*, vol. 36, p. 332, 1999.

[14] J. Caro, K. Harpole, V. Kasinath, et al.., *Proc. Natl. Acad. Sci. USA*, vol. 114, p. 6563, 2017.

[15] E. O'Brien, B. Fuglestad, H. Lessen, et al., *Angew. Chem. Int. Ed.*, vol. 59, p. 11108, 2020.

[16] B. Marques, M. Stetz, C. Jorge, K. Valentine and A. Wand, *Sci. Rep.*, vol. 10, p. 17587, 2020.

[17] R. Woldeyes, D. Sivak and J. Fraser, *Curr. Opin. Struct. Biol.*, vol. 28, p. 56, 2014.

[18] R. Barlow and K. Burston, *Brit. J. Pharmacol.*, vol. 66, p. 581, 1976.

6 Kinetics of Drug–Receptor Complex Formation

6.1 LIGAND BINDING KINETICS AND RESIDENCE TIME

The kinetics of ligand–receptor binding is described by the association and dissociation rate constants. The maximum (controlled by diffusion) bimolecular association rate constant, k_1, calculated using Equation (4.1) can be estimated using the Smoluchowski equation [1]:

$$k_1 = 4\pi R_{1,2} D_{1,2} \frac{N}{10^3} \quad \text{M}^{-1}\cdot\text{s}^{-1} \tag{6.1}$$

where $R_{1,2}$ is the sum of the radii of the active centers of the ligand and receptor, $D_{1,2}$ is the sum of their diffusion coefficients, and N is the Avogadro number.

If $R_1 = R_2$ is 5×10^{-8} cm, then D_2 is 0 for cell receptors immobilized onto the cell membrane and D_1 is 6×10^{-6} cm$^2\times$s^{-1} for sucrose in water; therefore, the k_1 value is $\sim5\times10^9$ M$^{-1}\times$s^{-1}. Using Equation (6.1), the theoretical maximum value for k_1 is of the order 10^{10} M$^{-1}\times$s^{-1}. The maximum k_1 value is calculated based on the assumption that each collision of molecules is effective, although this is unlikely in real cases. The association rate constant typically ranges from 10^6 to 10^7 M$^{-1}\times$s^{-1} for ligand binding to receptors and enzymes [2]. Steric barriers that occur during ligand–receptor complex formation as well as conformational adjustments of interacting molecules and displacement of hydrate shells reduce the theoretical value of the association rate constant. It is also important to consider delays due to ligand binding to nonspecific sites, especially *in vivo*.

On the contrary, the k_{-1} value varies widely across the range of 10^{-4}–10^{-2} s^{-1} for pharmacological substances and 10^2–10^4 s^{-1} for enzyme substrates.

As the rate of ligand association with protein targets is much higher than that of ligand dissociation, it is often simple and reliable to measure k_{-1} and to calculate k_1 using the equation $k_1 = \dfrac{k_{-1}}{K_d}$.

There are also parameters such as the average complex lifetime $t = \dfrac{1}{k_{-1}}$ and half-lifetime $t_{1/2} = \dfrac{0.69}{k_{-1}}$.

The pharmacokinetic lifetime of a compound in systemic circulation differs from these quantities and refers to the biochemical aspect, rather than the pharmacological aspect, of drug–receptor binding kinetics.

Nowadays, the lifetime of a ligand–receptor complex is usually mentioned as residence time. The importance of residence time determination and consideration in the drug discovery process is widely acknowledged (e.g., see [3] and references therein). Moreover, sometimes, residence time may be more useful for drug candidate optimization than binding affinity, particularly in case of multistage binding process. Furthermore, residence time can be considered a good predictor of drug efficacy [4]. The problem of understanding and predicting drug–target binding kinetics is thoroughly discussed in [5].

The estimation of the lifetime of ligand–receptor complexes or ligand residence time can also be useful for the following purposes:

- For use as a parameter in pharmacological mathematical models to predict the drug effect duration, as the residence time of a drug on the biotarget is as important as its pharmacokinetic metabolic rate to the durable pharmacological effect
- For evaluating the lifetime of short-lived complexes, including intermediates at various stages during ligand–protein interactions
- For filtering compounds in virtual screening pipelines, where common docking and scoring are not enough to separate the leads
- For selecting the method to separate bound and free fractions of the radiolabeled ligand because common techniques are long and problematic for short-lived complexes.

The most widely used computational methods for residence time estimation are based on molecular dynamics formalism (τ –RAMD, steered MD, metadynamics) [6, 7, 8, 9] and machine learning [10]. For molecular dynamics–based methods, a high-resolution protein structure is needed, while for machine learning methods, careful selection of structural and electronic features for system description and model training are required. In case of acceptable conditions, methods suffice well for the estimation of residence time within reasonable time.

However, before these methods were invented and elaborated, we tried to use a much simpler approach.

6.2 DRUG–RECEPTOR COMPLEX LIFETIME–AFFINITY RELATIONSHIP

The value of k_{-1} can, in principle, be evaluated using the K_d value. A linear dependence of $lg\,k_{-1}$ (that is, $-lg\,t$) on $lg\,K_d$ follows from Equation (4.1), provided the term $lg\,k_1$ is relatively constant because of its diffusion nature. Applying the logarithm to both sides of Equation (4.1) gives

$$\lg k_{-1} = -\lg t = \lg K_d + \lg k_1 \tag{6.2}$$

To test this hypothesis, equilibrium and kinetic constants of complex formation (100 pairs of K_d and t values) for 82 compounds were statistically analyzed, including receptor agonists and antagonists, enzyme substrates and inhibitors, and ion channel blockers [11]. If the value of any of the three constants (k_{-1}, k_{+1}, or K_d) was not available, then it is calculated from the known constants. The limitation for data selection is that the compound's molecular weight should be no more than 600 Da (drugs with small molecular mass). The reasons to confine the set of considered substances are also discussed.

Some binding constants were determined using the radiolabeling method, and others were measured through electrophysiological experiments and by relaxation kinetics methods.

The following linear regression equation was obtained in [11]:

$$, \qquad \lg t = \left(-6.89 \pm 0.23\right) + \left(-1.01 \pm 0.03\right) \lg K_d \tag{6.3}$$

where t is the mean lifetime of a complex (in seconds), and K_d is expressed in M. The correlation coefficient r = 0.962, the F-test value is 911.8 (significance level < 0.01), and the slope is close to the theoretically expected value of -1.

Values of t expressed in Table 6.1 related to the K_d constants are calculated using Equation (6.3). The lifetime of a complex does not depend on the concentration of free substance in a surrounding medium.

Evidently, if the K_d (K_M) constant lies in the range of 10^{-5}–10^{-3} M typical for enzyme–substrate complexes [2], the lifetime of complexes corresponds to the enzyme turnover number ($1/t$) from 10^2 to 10^4 s^{-1}.

Usually, the K_d values of medicines range from 10^{-10} to 10^{-7} M. As for short-lived complexes corresponding to a K_d value of $\sim 10^{-8}$ – 10^{-7} M estimated from Equation (6.3), the receptor blockade is attainable with increased doses. In the case of substances with a K_d value ranging from $\sim 10^{-10}$ to 10^{-9} M, the t value is much higher, whereas their doses are smaller, as a rule. At $K_d < 10^{-11}$ M (sub-picomolar range), the lifetime exceeds the half-period of a receptor or enzyme renewal due to protein synthesis, which varies from several hours to several days [12, 13]. These substances are referred to as nonequilibrium receptor blockers.

As shown in Table 6.1, no receptor can have a ligand-binding affinity of less than 0.01 pM; otherwise, it would be completely inactivated.

Switching-off receptors and enzymes from functioning for a long period (more than the protein's lifetime) should be compensated by their elevated synthesis, which is unprofitable for the organism. Thus, the upper (maximal) limit for drug affinity should be in the range of $K_d \approx 0.1 - 0.01$ nM. As shown in Table 6.2, substances with $K_d < 0.1$ nM cannot be regarded as medicines, and they mostly

TABLE 6.1
Lifetime of Ligand-Receptor Complexes (t) vs K_d as Calculated by Regression (6.3)

K_d, M	t, s	K_d, M	t
10^{-1}	$1.3 \cdot 10^{-6}$	10^{-8}	15 s
10^{-2}	$1.3 \cdot 10^{-5}$	10^{-9}	2.6 min
10^{-3}	$1.4 \cdot 10^{-4}$	10^{-10}	27 min
10^{-4}	$1.4 \cdot 10^{-3}$	10^{-11}	4.6 h
10^{-5}	$1.4 \cdot 10^{-2}$	10^{-12}	47 h
10^{-6}	$1.5 \cdot 10^{-1}$	10^{-13}	20 days
10^{-7}	1.5		

TABLE 6.2
Affinity of Drugs and Nonmedicinal Agents (Marked by *) to Specific Receptors

Substance	Receptor Type	K_d, nM	Reference
Ketanserin	5-HT$_2$	0.39	[14]
FK 1052*		0.01	[15]
Prazosin	α_1 adrenergic	0.31	[16]
Yohimbine	α_2 adrenergic	0.71	[17]
Atropine	mAChR	0.55	[18]
QNB*		0.06	[19]
Tubocurarine	nAChR	34	[20]
Epibatidine*		0.01	[21]
Haloperidol	D$_2$	0.5	[22]
Spiperone*		0.06	[22]
Clonazepam	GABA	0.75	[23]
Olanzapine	H$_1$	0.087	[24]
Melatonin	Melatonin	0.37	[25]
2-Iodomelatonin*		0.03	[26]
Fentanyl	μ-Opioid	0.45	[27]
Lofentanil*		0.023	[28]
Nifedipine	Ca^{2+}- channel	0.23	[29]

belong to the poison category. We can say that the toxicity of a drug is attributed to its very high dose or an excessive number of complexes, whereas the toxicity of ligands with superior affinity to targets is determined by the quantity and quality of its complexes, that is, their extraordinary strength.

6.3   POSSIBLE REASON FOR A WIDE CONFIDENCE INTERVAL IN THE REGRESSION MODEL

The confidence intervals for residence time in regression (Equation (6.3)) is wide because of a considerable difference in the diffusion constant k_{+1} for various substances and because of the number of contact points in a "ligand–bioacceptor" complex, which increases with the molecular weight as well as the linear size of a ligand. This might occur through the following mechanism of complex dissociation. Under the equilibrium state, the potential energy of interacting molecules equals to $E = E_e + RT$, where E_e is the interaction energy between molecules located at the equilibrium distance d_e, RT is the thermal molecular kinetic energy, R is the gas constant, and T is the absolute temperature. Heat oscillations of the molecules produce fluctuation of value E. Consequently, the distance between the oscillating molecules varies within an interval $d_1 \div d_2$ as depicted in Figure 6.1. As intermolecular oscillations are anharmonic, the average intermolecular distance d_o is greater than d_e. As estimated by Webb [30] using the empirical equation

$$d_o = d_e + \frac{0.5}{\sqrt{E_e - 1}} \tag{6.4}$$

the difference $d_o - d_e$ is in the range of 0.2–0.4 Å and $d_2 - d_e$ can be 0.8 Å for most types of interactions between the ligand groups and proteins (according to Webb's estimation [22], up to 1.2 Å).

The other approach to estimate the amplitude of molecular oscillations is based on the determination of the difference between the real size of the groups and the size of the cavity formed in solution because of heat motion. Because of these heat oscillations, the van der Waals radii obtained from X-ray data are overestimated. For most of the monovalent ions, the cavity radius in the aqueous medium is on average 30%–35% greater than the corresponding ion radius [30].

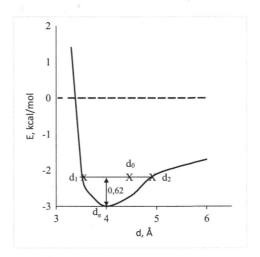

FIGURE 6.1 Typical curve for energy (E) dependence on time-averaged distance (d) between the interacting groups; the thermal kinetic energy is considered (based on data from [30]).

At certain moments, such oscillations are sufficient to significantly weaken intermolecular interactions. For example, the potential binding energy due to dipole–dipole and dispersion forces depends on the distance as $1/d^3$ and $1/d^6$, respectively. With increased distance between a phenyl moiety and the receptor surface, ranging from 3.8 Å (d_e) to 4.3 Å (as $\Delta d = 0.5$ Å), the binding energy is reduced by 1.7 kcal/mol and the lifetime of a ligand–receptor complex calculated using Equation (6.3) shortens by more than 10 times.

One might also consider that the distances between the interacting ligand and bioacceptor groups are changed asynchronously. Thus, the larger the size of the molecule, the greater is the number of contact points and the lower is the possibility of substantially worsened complex stability. If K_d values are close, then the lifetime of complexes, as a rule, becomes longer with increasing size, as well as increasing weight, of the molecules as demonstrated in Table 6.3. However, the greater weight of the molecules does not always coincide with the larger linear size, as it depends on molecular conformation in water.

At values closer to K_d, the residence time of peptide hormones or toxins on their specific receptors is higher than that of drug–receptor complexes. The higher the molecular weight, the lower the values of both k_{-1} and k_{+1} constants; therefore, an explicit dependence of K_d on molecule size is not observed. A decrease in the k_{+1} value is also due to the increased contact points, resulting in the necessity of longer time for the orientation of a molecule on the receptor and for conformational arrangement of interacting molecules[1]. As with the comparatively low-molecular triiodothyronine and progesterone molecules, the lifetime of their complexes is close to calculated values.

However, some deviations from the proposed lifetime dependence on molecular weight are also observed. Thus, glucagon and thyrotropin have the same kinetic characteristics as those of low-molecular-weight compounds. The reason can be that, apart from the number of contact points, the value of k_{-1} and complex lifetime are influenced by the strength of forces that fix a molecule on the protein.

TABLE 6.3
Binding Constants of Hormones and Toxins to Specific Receptors

Substance	Molecular weight	Tissue	K_d, M	k_{+1}, $M^{-1}s^{-1}$	k_{-1}, s^{-1}	$t = \dfrac{1}{k_{-1}}$	t_{model} *	Ref.
Lutropin	~$3\cdot10^4$	Rat Leidig cells	$(1–2)\cdot10^{-10}$	$1\cdot10^4$	$2\cdot10^{-6}$	139 hours	18 min	[31]
Gonadotropin	~$3\cdot10^4$	Rat luteal cells	$1.7\cdot10^{-10}$	$2.8\cdot10^6$	$2.1\cdot10^{-5}$	13 hours	16 min	[32]
Thyrotropin	$2.8\cdot10^4$	Thyroid cells	$5.3\cdot10^{-10}$	$1\cdot10^7$	$6\cdot10^{-3}$	2.8 min	5.0 min	[33]
Nerve growth factor	$1.3\cdot10^4$	Bovine brain	$3.0\cdot10^{-9}$	$2.3\cdot10^5$	$7.0\cdot10^{-4}$	24 min	52 s	[2]
Naja haje α-toxin	$7.8\cdot10^3$	Torpedo electric organ membrane	$6.7\cdot10^{-10}$	$4.8\cdot10^4$	$3.2\cdot10^{-5}$	8.7 hours	4.0 min	[34]
α-bungarotoxin	$(6–7)\cdot10^3$	Eel electrophorus organ	$(1–2)\cdot10^{-7}$	$1.2\cdot10^4$	$1.8\cdot10^{-3}$	9.3 min	1.0 s	[35]
Insulin	$5.8\cdot10^3$	Rat liver	$8.3\cdot10^{-9}$	$(0.13–1)\cdot10^6$	$7\cdot10^{-4}$	24 min	19 s	[36]
		Rat fat cells	$3.0\cdot10^{-9}$	$4.1\cdot10^5$	$1.1\cdot10^{-3}$	15 min	52 s	[37]
		Rat liver	$1.6\cdot10^{-10}$	$2.3\cdot10^6$	$3.8\cdot10^{-4}$	44 min	17 min	[38]
Glucagon	$3.5\cdot10^3$	Rat liver	$6.7\cdot10^{-10}$	$1\cdot10^6$	$4\cdot10^{-3}$	4.2 min	4.0 min	[36]
Somatostatin	$1.6\cdot10^3$	Rat lung	$1.2\cdot10^{-10}$	$1\cdot10^7$	$1.2\cdot10^{-4}$	2.3 hours	23 min	[39]
Vasopressin	$1.1\cdot10^3$	Pig kidney	$2\cdot10^{-9}$	$3.8\cdot10^5$	$7.5\cdot10^{-4}$	22 min	1.3 min	[40]
Angiotensin	~$1\cdot10^3$	Rat adrenal cortex	$2.1\cdot10^{-9}$	$2.4\cdot10^5$	$5\cdot10^{-4}$	33 min	1.3 min	[41]
Triiodothyronine	651	Rat liver nuclei	$1.6\cdot10^{-10}$	$5.8\cdot10^6$	$9.6\cdot10^{-4}$	17 min	17 min	[42]
Progesterone	314	Mouse oviduct	$2\cdot10^{-10}$	$1.2\cdot10^7$	$2.3\cdot10^{-3}$	7.2 min	13 min	[43]

* Estimated using Equation (6.3) from the K_d value.

Significant fluctuations around the average energy E_e in the ligand–protein complex explain a difference between the mechanism of the breakdown of ligand–protein complexes and chemical bonds breaking between a protein and a covalently bound ligand, even if the total energy of many nonvalent interactions is comparable to the energy of a covalent bond. The dissociation rate of the complex can be increased by the competitive displacement of one ligand with another, whereas a covalently bound ligand cannot be dissociated in such a way.

6.4   STAGES OF LIGAND BINDING NOT CONTROLLED BY DIFFUSION

In Chapter 4, we have considered the first stage in ligand–receptor interaction (Scheme 4.1), which is often final. However, the other scenario is possible: the formation of a primary complex does not yield the equilibrium state and is followed by its further transformation by three possible pathways (Schemes 4.5–4.7).

Upon binding to receptors, the equilibrium state is rapidly achieved for certain ligands, while the duration is long for other substances, as evident from Table 6.4.

Based on the onset of the equilibrium state within 5–10 min, it can be assumed that dicyclomine and pirenzepine interact with mAChR by a single-stage pattern. As for the other substances listed in Table 6.4, the first (fast) stage is followed by a stage when the K_d values slowly decline for N-methyl-scopolamine (NMS), scopolamine, and dexetimide and, vice versa, slowly grow for trihexyphenidyl and AF-DX116.

Binding of N-methyl-4-piperidyl benzilate [45], 3-quinuclidinyl benzilate (QNB) [46], and some other substances to mAChR involves two stages. The second stage involves a relatively slow transformation of a ligand–receptor complex to the state that binds the ligand more firmly, and Kloog and Sokolovsky interpreted it as receptor "isomerization" [45]. In the meantime, a two-site model of the antagonist NMS (Figure 6.2) binding to mAChR was proposed [47], where the ligand first binds to the peripheral site of the receptor and then binds to the orthosteric site with much stronger affinity. The model was further confirmed to show NMS allosteric binding (Scheme 4.6) and interaction with residues on the extracellular loops of the receptor [48]. However, QNB two-staged binding was not considered.

Indeed, a few substances are characterized by varying binding constants such as QNB (Figure 6.2). The K_d constant of QNB binding to the rat brain mAChR is measured as $5.7 \cdot 10^{-10}$ M [49] or $(1–2) \cdot 10^{-12}$ M [50], and half-decay time of the formed complex is determined as 57 min [19] or 10 h [50]. Nevertheless, strong QNB binding to mAChR is not irreversible: [3H]-QNB is shown to

---

TABLE 6.4
Changes in the K_d Values of Muscarinic Acetylcholine Receptor (mAChR) Antagonists During Their Incubation with Rat Pancreatic Homogenate [44]. Ligands' Structure Is Depicted in Figure 6.2

mAChR Antagonists	K_d, nM						
	5 min	10 min	20 min	40 min	60 min	120 min	240 min
[3H] - NMS	1.4	0.62	0.33	0.20	0.15	0.13	0.10
Scopolamine	2.5	1.0	0.60	0.40	0.30	0.27	0.24
Dexetimide	20	6	4	2	1.1	0.7	0.4
Dicyclomine	15	11	10	9	9	10	8
Trihexyphenidyl	13	10	10	13	15	20	18
Pirenzepine	350	190	180	200	190	180	190
AF-DX116	1300	1000	1200	1600	1800	1900	2100

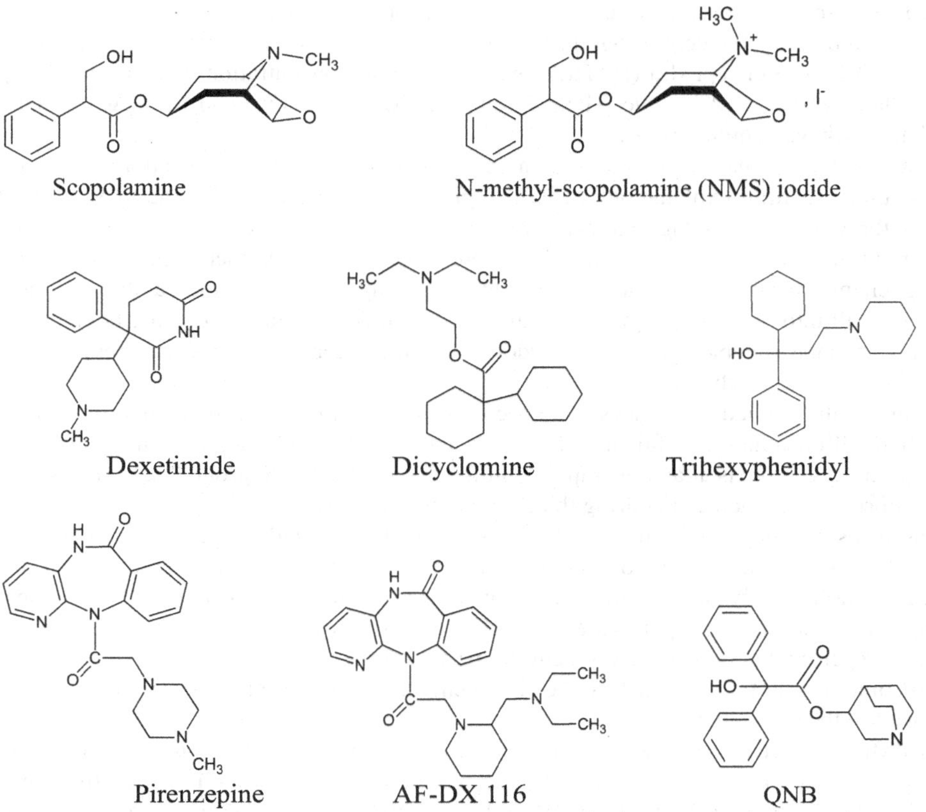

Scopolamine

N-methyl-scopolamine (NMS) iodide

Dexetimide

Dicyclomine

Trihexyphenidyl

Pirenzepine

AF-DX 116

QNB

FIGURE 6.2 mAChR antagonists.

be removed from complexes by unlabeled antagonists at high concentrations [19, 51]. Noteworthy, only L-QNB induces the so-called isomerization of the complex, but not D-QNB, which binds according to the single-stage scheme [52]. As the isomerization of a primary complex is a very long stage, [³H]-QNB was regarded to be unsuitable for measurements under the equilibrium conditions, especially for the determination of the K_d values for unlabeled agonists and antagonists using the competitive displacement method [50]. Furthermore, QNB is not selective to various subtypes of muscarinic acetylcholine receptors [53].

As the second stage is not observed for all substances, it would be useful to discuss the underlying molecular mechanisms. Several explanations beyond the mechanism proposed in [45, 48] can be considered while not being mutually exclusive.

Along with fast and small (0.1 Å) heat fluctuations of atoms or group of atoms, slow motions of protein domains at higher amplitudes and periods (up to milliseconds) occur [54], and a primary ligand–receptor complex can be rearranged in a stepwise manner. These rearrangements may strengthen one or several hydrogen bonds and extend the ligand–receptor contact area. Some chemical groups in a ligand molecule are on the receptor surface at the beginning of complex formation, and they might come into closer contact with side chains of the receptor's binding site residues. This could improve the stability of the complex.

Another explanation is also possible. Spontaneous conformational changes in the receptor protein reduce complementarity and the ligand affinity falls. Structural oscillations on ligand binding in proteins are reversible, but it may be assumed that the protein relaxation in a local binding site is not attained because of a stabilizing effect of a ligand–receptor complex. The appearance of oscillations during ligand binding to receptors has already illustrated in [55] for the δ-opiate receptor ligand [³H]-[D-Ala-2, D-Leu-5]-enkephalin (DADLE) when its bound concentration does not reach a plateau and oscillates within a 50% range. The phenomena, observed in [55] and for mAChR, are not the same but may have common roots.

Notwithstanding that fluctuations can be invoked to explain ligand–receptor complex isomerization, such an explanation sheds no light on the reasons why some substances bind to mAChR through a single-stage pathway and other structurally similar ligands interact with this receptor through a two-stage scheme. Therefore, we consider another explanation of different binding schemes for the muscarinic antagonists. According to the experimental data, all the studied substances fall into two groups: group 1 (Table 6.5) includes substances that are bound to mAChR through one-stage scheme; group 2 includes substances (Table 6.6) that are bound to mAChR through the two-stage scheme.

Sorting of the studied substances to a specific group is based on experimental data on mAChR from rat small intestines. As shown in Table 6.6, the equilibrium between group 1 substances and acetylcholine receptors is attainable rapidly, unlike that attained with group 2 substances, for which the equilibrium is not achieved during the experiment timescale.

Substances of group 1 at high concentrations are washed off and replaced by acetylcholine even after prolonged incubation. By contrast, ligands belonging to group 2 are removed and displaced with acetylcholine only after short-term incubation (2–5 min) and at low concentrations. After prolonged incubation, these ligands are not washed off, even with high acetylcholine concentrations.

The long period that is necessary to achieve the steady state (5–7 min) at the first stage of ligand–mAChR binding can be explained by the slow penetration of substances through the rat intestinal tissue wall in the experiments.

The difference between the two groups is also evident from experiments *in vivo* [56]. Thus, compound C-**5.1** from group 1 is present in the rat brain tissues until it is eliminated from the blood, and only its trace amounts are found in the rat brain within 6 h after the injection. The time of half-disappearance of compound C-**5.1** from the rat brain ($t_{1/2}$) is less than 1 h.

On the contrary, [¹⁴C]-QNB behaves as a nonequilibrium antagonist, and its complex formation with mAChR from the rat brain is stable during several days in the absence of free substance in the blood. Extrapolation of data on the QNB content in the brain gives the $t_{1/2}$ value of approximately 10 days. The data on the elimination of [³H]-QNB and compound C-**6.17** from the rat brain [56] show that their $t_{1/2}$ values are 3.5–4 days and 8–9 days, respectively.

There are specific structural differences of ligands belonging to group 1 and group 2. All substances of group 2 contain a quinuclidyl moiety as well as hydroxy or methoxy groups at the α-C atom in the acidic group. However, these structural features are necessary but not sufficient to attribute the substances to the second group. For example, compounds C-**5.5** and C-**6.8** belong to group 1 as well as (S)-isomers of group 2 substances. As compounds C-**5.1** and C-**6.10** belong to group 1, we suggest that the hydroxy or methoxy group plays a main role in the specific binding mechanism of QNB and QNB-like substances.

The possible explanations for the two-stage binding of group 2 substances can be suggested:

(i) Ligands of group 2 contain a hydroxy or methoxy group, which forms a strong hydrogen bond at the second stage of interaction with the receptor protein. It is possible that the hydroxy group of the compounds from group 1 and group 2 forms hydrogen bonds with different receptor residues. It follows from different partial contributions of the hydroxy group to the binding free energy of a primary complex between various ligands and mAChR. The difference between equilibrium and nonequilibrium antagonists can be seen by comparing Table 6.7 and Table 6.8.

TABLE 6.5
mAChR Antagonists with One-Stage Binding Scheme [56]

Compound	K_d, nM	Compound	K_d, nM
Atropine	0.36	C-5.5	2.2
Benactyzine	2.90	(S)-C-6.9	0.83
Spasmolytin	34	C-6.10	0.092
(S) – QNB	0.58	C-4.11	0.056
C-5.1	0.20	C-4.42	0.61
C-6.8	1.8	C-4.17	0.29

TABLE 6.6
mAChR Antagonists with Two-Stage Binding Scheme [56]

Antagonist	K_d, nM	Antagonist	K_d, nM
QNB	0.14	(R,R)-C-6.14	0.092
(R) - QNB	0.068	C-6.15	0.11
C-6.12	6.8	(R)-C-6.16	0.047
C-6.13	0.29	(R,R)-C-6.17	0.14
C-5.4	0.29		

TABLE 6.7
IC$_{50}$ for Acetylcholine-Induced (5.5·10^{-6} M) Effect with Incubation Time [56]

Group	Compound	Optical Isomer	IC$_{50}$, M Incubation Time, min					
			2	5	10	20	30	60
1	C-4.11	Achiral	5.6·10^{-9}	3.1·10^{-9}	3.0·10^{-9}	—	—	—
	C-5.5	Racemate	1.3·10^{-8}	1.1·10^{-8}	1.4·10^{-8}	1.2·10^{-8}	—	—
	Atropine	Racemate	2.8·10^{-8}	2.0·10^{-8}	—	2.0·10^{-8}	—	—
	Benactyzine	Achiral	2.6·10^{-7}	1.6·10^{-7}	1.5·10^{-7}	1.7·10^{-7}	—	—
2	QNB	Racemate	1.0·10^{-8}	7.8·10^{-9}	3.7·10^{-9}	1.1·10^{-9}	—	6.9·10^{-10}
	QNB	(R)	7.5·10^{-9}	3.8·10^{-9}	1.3·10^{-9}	5.1·10^{-10}	3.0·10^{-10}	—
	C-6.13	Racemate	—	1.6·10^{-8}	9.1·10^{-9}	5.5·10^{-9}	—	—
	C-6.14	(R, R)	5.1·10^{-9}	2.8·10^{-9}	1.4·10^{-9}	8.1·10^{-10}	—	2.4·10^{-10}
	C-6.15	Racemate	7.7·10^{-9}	6.1·10^{-9}	2.0·10^{-9}	5.0·10^{-10}	—	—

TABLE 6.8
Partial Contribution of a Hydroxyl Group to the Free Energy of Complex Formation with mAChR [56]

Compound	K$_d$, nM	ΔG, kcal/mol	ΔΔG, kcal/mol
C-4.11	0.056	−14.6	
C-4.24	0.61	−13.1	−1.5
Benactyzine	2.9	−12.2	
Adiphenine	3.4	−10.6	−1.6

TABLE 6.9
Partial Contribution from Hydroxyl Group to the Binding Free Energy ($\Delta\Delta G$)
for Substances of Group 2 with mAChR [56]

Ligand	K_d, nM	ΔG, kcal/mol	$\Delta\Delta G$, kcal/mol
QNB	0.14	−14.0	
C-5.1	0.20	−13.8	−0.2
C-6.15	0.11	−14.2	
C-6.10	0.092	−14.3	0.1

The contribution of the hydroxyl group in antagonists in equilibrium is estimated as 1.5–1.6 kcal/mol (Table 6.7), and this value is comparable to the hydrogen bonding energy with the receptor. However, as implied from Table 6.8, at the first stage during the binding of QNB or compound C-**6.15**, a hydrogen bond is not formed. The slow formation rate of hydrogen bonds by nonequilibrium antagonists can be explained by slow oscillations in structural fragments of the receptor protein. If this mechanism of hydrogen bonding occurs, the appropriate conditions for this interaction with different muscarinic cholinergic receptors appear asynchronously.

(ii) An interaction between substances of group 2 and mAChR is not terminated by hydrogen bonding. It may involve hydrogen atom transfer from the receptor's proton donor residue such as tyrosine [57] to the hydroxy group of the oxygen atom with heterolysis of a hydroxyl-carbon bond, resulting in the formation of a carbocation and water molecule as the leaving group, as depicted in Figure 6.3.

Proton transfer results in an additional ion–ion interaction between the ligand and receptor. Incomplete reaction without the formation of a covalent bond can be explained by steric hindrances possessed by both the structural fragments of the ligand and receptor molecules and a water molecule. A computational model [58] for the HOH ... $^-$OH system demonstrated that, by decreasing the distance between oxygen atoms d_{O-O} to 2.7 Å, the proton localizes either

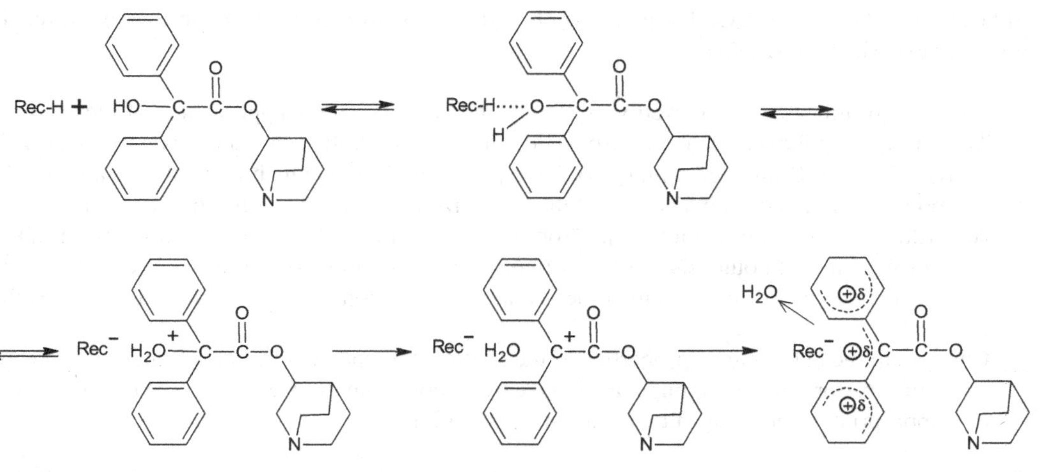

FIGURE 6.3   Hypothetical scheme of the interaction of the QNB hydroxy group with its receptor.

near one or more oxygen atoms. A potential energy barrier between two equilibrium states of proton was estimated as 5 kcal/mol. Fluctuations of the distance d_{0-0} may be advantageous for proton transfer. In more complex systems, as noted in [58], nonvalent interactions between molecular fragments may be sufficient for rapid proton transfer. For the HOH ... $^-$OH system in an aqueous medium, the data [59] indicate that the free energy barrier height for proton transfer might be lowered to 0.5 kcal/mol. The hydroxyl group of substances with more flexible conformation than that of QNB, for example, ligand C-**4.11** or benactyzine (Figure 5.1), may take the same position on the receptor as the hydroxyl group of QNB does. However, the hydroxyl group in ligand C-**4.11** and benactyzine is likely to occupy the position that is more thermodynamically favorable for primary complex formation through hydrogen bonding. It is worth to noting that the quinuclidyl radical consists of three six-membered rings. Perhaps this difference between the cationic groups of QNB and other ligands affects the position of the former on the receptor.

The formation of carbocations from benzylic acid in strongly acidic solutions was investigated in [60]. In neutral and weakly acidic solutions, these carbocations are formed; the conditions of their formation may occur only on the receptor because of "forced" approaching of the corresponding groups at low dielectric constants. Benzylic carbocation is stable due to charge delocalization in benzene rings.

The energy increment during the slow nondiffusion stage of QNB binding to muscarinic ACh receptors can approximately be estimated from the ratio of half-life times ($t_{1/2}$) of complexes formed at the first and second stages. The $t_{1/2}$ value of the complex formed at the second stage of QNB–receptor interaction can be derived from the data of *in vivo* experiments that allow the extension of the observation time and to exclude certain errors due to spontaneous denaturation of receptor proteins *in vitro*. The half-life time ($t_{1/2}$) of mAChR is approximately 20 h at 35°C [61]. From the experiments with radiolabeled ligands [56], it follows that the half-life of the complex ($t_{1/2}$) with |^{14}C|-labeled QNB is approximately 10 days, while for the complex with |^{3}H|-labeled QNB a $t_{1/2}$ value ranges from 3.5 to 4 days. Regarding the mean $t_{1/2}$ value of 7 days, the ratio of $t_{1/2}$ values for the complexes formed at the first and second stages is approximately 1000, which corresponds to the binding free energy difference, ΔG, equivalent to 4–4.5 kcal/mol.

Even so, based on the 4–4.5 kcal/mol increase in the binding energy, we cannot choose one of the two suggested mechanisms of QNB binding at the second stage. This energy increment can be ascribed to either strong hydrogen bonding or an interionic interaction. In our opinion, the second explanation to be used for QNB binding with carbocation formation is suitable to interpret the following experimental results:

(1) The equilibrium state is achieved very slowly and depends on the approaching of ions formed. The onset of equilibrium may be associated with the slow translocation or displacement of the "leaving" group. Compared with QNB, the equilibrium state is much more slowly attained for compound C-**6.12**, possibly because of the large size of its "leaving" group (CH_3OH).

(2) According to the experiments [56], atropine (Figure 4.2) behaves as an equilibrium antagonist of mAChR, although its molecule contains a rigid amino group. It has been known that substances with a hydroxyl group at the primary carbon atom are not good producers of stable carbocation.

(3) Changes in the antagonistic properties of the studied compounds with substitution of both aromatic groups for aliphatic groups (as in the case of compound C-**5.5**, Table 6.5) may occur due to inappropriate conditions of carbocation stabilization.

The ligand–receptor complex formation, which is represented as a two-staged process, may, in fact, involve many stages and numerous forms of complexes. Because of the relatively large concentrations of acetylcholine used as a competitive test compound, the experimental methods may not reveal the second stage of complex formation for amino glycolates with affinities that change insignificantly as time progresses. However, radioligand binding assays are useful to demonstrate the peculiar pattern of the series of quinuclidyl glycolates binding to mAChR, involving the formation of a uniquely strong complex. Interestingly, this complex is stable through the entire duration of experiments. Thus far, the time scale of equilibrium state achievement for QNB has not been established.

It is a common viewpoint in molecular pharmacology that chemical reactions between a ligand and a protein result in covalent bonding, as appears as the final effect of mercurial and arsenical organic compounds that bind the -SH cysteine groups of proteins, as well as agents that phosphorylate or alkylate enzymes or receptors. The common group in the substances' structure responsible for alkylation of the receptor's anionic site is N-β-chloroethyl. The mechanism of secondary complex formation through carbonium cation (if it occurs) has the particular interest itself, as the chemical reaction between the receptor and ligand results in the formation of a complex (not a chemical compound) and its lifetime might be controlled through structural modifications of ligands. In this regard, it is worth mentioning the novel concept of photo-switchable ligands for mAChR proposed in [62]. The authors synthesized a ligand from two fragments with affinity to orthosteric and allosteric binding site as well as an azobenzene linker. The produced molecule changes its conformation upon laser emission, thus controlling orthosteric binding.

Thus, time-dependent transformation of receptor binding properties upon ligand action, known as the "receptor isomerization," may proceed through clear molecular mechanisms.

NOTE

1) A decrease in the k_{+1} value with increasing molecular weight is not directly related to a lowering diffusion coefficient that is inversely proportional to the molecular radius. As for spherical molecules, an increment in the molecular weight from 600 to 5000 produces only a twofold enlargement of the molecular radius.

REFERENCES

[1] M.V. Smoluchowski, *Phys. Z.*, vol. 17, p. 557, 1916.
[2] S. Varfolomeev and K. Gurevich, Biokinetics: A Practical Guide (in Russian), Moscow: Grand, 1999, p. 348.
[3] R. Copeland, *Nat. Rev. Drug Discov.*, vol. 15, p. 87, 2016.

[4] M. Bernetti, M. Masetti, W. Rocchia and A. Cavalli, *Ann. Rev. Phys. Chem.*, vol. 70, p. 143, 2019.

[5] H. Lu, J. Iuliano and P. Tonge, *Curr. Opin. Chem. Biol.*, vol. 44, p. 101, 2018.

[6] A. Nunes-Alves, D. Kokh and R. Wade, *Curr. Opin. Struct. Biol.*, vol. 64, p. 126, 2020.

[7] N. Bruce, G. Ganotra, S. Richter and R. Wade, *J. Chem. Inf. Model.*, vol. 59, p. 3630, 2019.

[8] J. Patel, A. Berteotti, S. Ronsisvalle, W. Rocchia and A. Cavalli, *J. Chem. Inf. Model.*, vol. 54, p. 470, 2014.

[9] H. Sun, Y. Li, M. Shen, D. Li, Y. Kang and T. Hou, *J. Chem. Inf. Model.*, vol. 57, p. 1895, 2017.

[10] A. Potterton, A. Heifez and A. Townsend-Nicholson, Artificial Intelligence in Drug Design. Methods in Molecular Biology, vol. 2390, p. 191, 2022.

[11] F. Dukhovich, E. Gorbatova, M. Darhovskii and V. Kurochkin, *Pharm. Chem. J.*, vol. 36, p. 248, 2002.

[12] B. Alberts, D. Bray, J. Lewis, M. Raff, K. Roberts and J. Watson, Molecular Biology of the Cell, New York: Garland Science, 1994.

[13] R. Marri, D. Grenner, P. Meyes and V. Rodwell, Human Biochemistry, vol. 1, Moscow: Mir, 1993, p. 103 (in Russian).

[14] J. Leysen, C. Niemegeers, J. Van Nueten and P. Laduron, *Mol. Pharmacol.*, vol. 21, p. 301, 1988.

[15] Y. Nagakura, M. Kadowaki, K. Tokoro, M. Tomoi, J. Mori and M. Kohsaka, *J. Pharmacol. Exp. Ther.*, vol. 262, p. 752, 1993.

[16] J. Doxey, A. Lane, A. Roach and N. Virdee, *Naunyn Schmiedebergs Arch. Pharmacol.*, vol. 325, p. 136, 1984.

[17] M. Owens, W. Morgan, S. Plott and C. Nemeroff, *J. Pharmacol. Exp. Ther.*, vol. 283, p. 1305, 1997.

[18] M. McKinney, D. Anderson, C. Forray and E. el-Fakahany, *J. Pharmacol. Exp. Ther.*, vol. 250, p. 565, 1989.

[19] H. Yamamura and S. Snyder, *Proc. Nat. Acad. Sci. USA, Biol. Sci.*, vol. 71, p. 1725, 1974.

[20] P. Asher, W. Large and H. Rang, *J. Physiol. (Gr. Brit.)*, vol. 295, p. 139, 1979.

[21] M. Marks, K. Smith and A. Collins, *J. Pharmacol. Exp. Ther.*, vol. 285, p. 377, 1998.

[22] B. Roth, S. Tandra, L. Burgess, D. Sibley and H. Meltzer, *Psychopharmacology (Berl.)*, vol. 120, p. 365, 1995.

[23] M. Massotti, J. Schlichting, M. Antonacci, P. Giusti, M. Memo, E. Costa and A. Guidotti, *J. Pharmacol. Exp. Ther.*, vol. 256, p. 1154, 1991.

[24] E. Richelson and T. Souder, *Life Sci.*, vol. 68, p. 29, 2000.

[25] M. Dubocovich, M. Masana, S. Iacob and D. Sauri, *Naunyn Schmiedebergs Arch. Pharmacol.*, vol. 355, p. 365, 1997.

[26] I. Beresford, C. Browning, S. Starkey, et al., *J. Pharmacol. Exp. Ther.*, vol. 285, p. 1239, 1998.

[27] I. Goldberg, G. Rossi, S. Letchworth, et al., *J. Pharmacol. Exp. Ther.*, vol. 286, p. 1007, 1998.

[28] P. Maguire, N. Tsai, J. Kamal, C. Cometta-Morini, C. Upton and G. Loew, *Eur. J. Pharmacol.*, vol. 213, no. 2, p. 219, 1992.

[29] A. Schotte, P. Janssen, W. Gommeren, W. Luyten, P. Van Gompel, A. Lesage, K. De Loore and J. Leysen, *Psychopharmacology (Berl.)*, vol. 124, p. 57, 1996.

[30] J. Webb, Enzyme and Metabolic Inhibitors, vol. 1, New York: Academic Press, 1966, p. 237 (in Russian).

[31] K. Catt and M. Dufau, *Nat. New Biol.*, vol. 244, p. 219, 1973.

[32] C. Rao and B. Saxena, *Biochim. Biophys. Acta*, vol. 313, p. 372, 1973.

[33] B. Verrier, G. Fayet and S. Lissitzkys, *Eur. J. Biochem.*, vol. 42, p. 355, 1974.

[34] R. Chicheportiche, J. Vincent, C. Kopeyan, H. Schweitz and M. Lazdunski, *Biochemistry*, vol. 14, p. 2081, 1975.

[35] G. Hess, J. Bulger, J. Fu, E. Hindy and R. Silberstein, *Biophys. Res. Commun.*, vol. 64, p. 1018, 1975.

[36] P. Freychet, G. Rosselin, F. Rancon, M. Foucherean and Y. Broer, *Horm. Metab. Res.*, vol. 5, p. 72, 1974.

[37] S. Glieman, S. Gammeltoft and J. Vinten, *J. Biol. Chem.*, vol. 250, p. 3368, 1975.

[38] S. Jacobs and P. Cuatrecasas, In: *Hormone-Receptor Interactions, Molecular Aspects*, New York: Marcel Dekker, 1976.

[39] J. Schloos, F. Raulf, D. Hoyer and C. Bruns, *Brit. J. Pharmacol.*, vol. 121, p. 963, 1997.

[40] J. Bockaert, C. Roy, R. Rajerison and S. Jard, *J. Biol. Chem.*, vol. 248, p. 5922, 1973.

[41] H. Glossman, A. Baukal and K. Catt, *J. Biol. Chem.*, vol. 249, p. 825, 1974.

[42] M. Surks and J. Oppenheimer, In: *Hormone-Receptor Interaction, Molecular Aspects*, New York: Marcel Dekker, 1976.

[43] G. Tkacheva, M. Balabolkin and I. Laricheva, Radio-Immune-Chemical Experimental Methods, Moscow: Medicine, 1983, pp. (a) p. 130, (b) p. 115 (in Russian).

[44] M. Waelbroeck, J. Camus and J. Christophe, *Mol. Pharmacol.*, vol. 36, p. 405, 1989.

[45] I. Kloog and M. Sokolovsky, *Biochem. Biophys. Res. Commun.*, vol. 81, p. 710, 1978.

[46] J. Järv, B. Hedlund and T. Bartfai, *J. Biol. Chem.*, vol. 254, p. 5595, 1979.

[47] J. Jakubík, E. El-Fakahany and S. Tuček, *J. Biol. Chem.*, vol. 275, p. 18836, 2000.

[48] J. Jakubík, A. Randáková, P. Zimčik, E. El-Fakahany and V. Doležal, *Sci. Rep.*, vol. 7, p. 40381, 2017.

[49] P. Moingeon, J. Bidart, G. Alberisi, C. Boudene and C. Bohuon, *Toxicology*, vol. 31, p. 135, 1984.

[50] M. Eller, A. Kyiv, U. Langel, A. Rinken and Y. Yarv, *Neurokhimiya*, vol. 6, no. 4, p. 605 (in Russian), 1987.

[51] M. Schimerlik and R. Searcles, *Biochemistry*, vol. 19, no. 15, p. 3408, 1980.

[52] T. Gaginella, T. Rimele, T. O'Dorisio and R. Dorh, *Life Sci.*, vol. 26, p. 1599, 1980.

[53] S.-H. Gilani and L. Cobbin, *Trends. Pharmacol. Sci.*, vol. 9, p. 87, 1988.

[54] A. Rubin, Fundamentals of Biophysics, New York: Wiley, 2014, p. 121.

[55] T. Sukhomlin, E. Melikhova, I. Kurochkin and S. Varfolomeev, *Biochemistry (Moscow)*, vol. 57, p. 1142, 1992.

[56] F. Dukhovich, M. Darkhovskii, E. Gorbatova and V. Kurochkin, Molecular Recognition: Pharmacological Aspects, New York: Nova Science Publishers, 2004.

[57] A. Nilsen-Moe, C. Reinhardt, S. Glover, L. Liang, S. Hammes-Schiffer, L. Hammarström and C. Tommos, *J. Am. Chem. Soc.*, vol. 142, p. 11550, 2020.

[58] K. Burshtein and Y. Khurgin, *Russ. Chem. Bull.*, vol. 23, p. 266, 1974.

[59] N. Agmon, H. Bakker, R. Campen, R. Henchman, P. Pohl, S. Roke, M. Thamer and A. Hassanali, *Chem. Rev.*, vol. 116, p. 7642, 2016.

[60] H. Dobeneck and R. Kiefer, *Ann. Chem.*, vol. 684, p. 115, 1965.

[61] Y. Langel, A. Rinken, L. Tyakhepyld and J. Järv, *Neurochemistry*, vol. 1, p. 343, 1982 (in Russian).

[62] L. Agnetta, M. Kauk, M. Canizal, R. Messerer, U. Holzgrabe, C. Hoffmann and M. Decker, *Angew. Chem. Int. Ed.*, vol. 56, p. 7282, 2017.

7 Distant and Contact Interaction of Drugs and Receptors

7.1 WHAT IS THE MAJOR PROBLEM IN UNDERSTANDING MOLECULAR RECOGNITION?

The term "recognition" refers to the formation of a stable complex by a substance with certain type(s) of biomolecules.

Recognition of native receptors implies the complementary "lock-and-key" fit (mirror-symmetric similarity) between interacting molecules or active (binding) sites in biomolecules. The major problem is to understand how ligand molecules escape from trapping by numerous and diverse nonspecific acceptors, not to decipher how a ligand binds to a specific target. Explanations for molecular recognition are based on the following principle: "two molecules with complementary surfaces tend to join together and interact, whereas molecules without complementary surfaces do not interact" [1]. This explanation is inaccurate because noncomplementary molecules also form complexes. Complex formation between molecules without complementarity is driven by universally present dispersion interactions. To demonstrate, consider that drug molecules usually contain several anchoring groups such as phenyl, cyclopentyl, cyclohexyl, thienyl, and others. The potential binding energy of phenyl to an arbitrary protein can be estimated as ~3 kcal/mol for one-side contact; if phenyl is immersed into a protein hydrophobic cavity or cell membrane, then the binding energy can reach ~9 kcal/mol [2]. These estimations are consistent with the experimental heats of sorption of benzene, methyl acetate, cyclopentyl, and cyclohexyl on graphite black that resembles the hydrophobic surface: 9.42, 7.4, 7.06, and 7.59 kcal/mol, respectively [3].

Relatively strong sorption of substances on graphite black is due to the binding effect of all parts in the carbon lattice. Biomolecules contain a variety of polar groups that may bind or repulse numerous molecules. All or almost all interactions between complementary particles only participate in binding.

One may suggest that the discussed nonspecific sorption negatively influences the binding of drugs applied with low doses by increasing its latent period of action. Atropine, clonidine, lisuride, gynipral, ethinyl estradiol, and some other well-known drugs are effective at doses of 0.01–0.1 mg [4]. These doses correspond to average concentrations in the range of 10^{-9}–10^{-8} M in the organism, which is comparable to the average concentration of specific receptors in mammals [5, 6]. Nonetheless, such drugs bind to their specific receptor during a usual latent period. How long may be the lifetime of nonspecific binding to avoid substantial capture on nonspecific acceptors? For this, estimation of the amounts of specific and nonspecific binding sites for each drug molecule is needed.

DOI: 10.1201/9781003366669-8

7.2 PROPORTION BETWEEN SPECIFIC AND NONSPECIFIC ACCEPTORS OF DRUGS

Assessment of the ratio between the number of nonspecific acceptors and the number of specific binding sites can be carried out in the following manner. By the order of magnitude, this value is defined as the ratio of the area of all possible binding places in the body to the area of binding sites of specific receptors for a certain drug substance. Each specific binding site for drugs with molecular weight of <600 has an area of ~1–2 nm^2, and the average receptor concentration of a certain type in mammals is 10^{-9}–10^{-8} M [5, 6, 7, 8]. Bearing in mind the Avogadro number, the area of functional receptors' binding sites is S_{bind} ~10^{13} nm^2.

Given the average size of animal somatic cells (1–2·10^4 nm in diameter), the overall cell surface per 1 g of the tissue can be taken as S_{cell} ~10^{18} nm^2. This is a good estimate: for example, the overall cell surface per 1 g of frog heart tissue is 6000 cm^2 (6·10^{17} nm^2) [9]. These values are applicable for the human body, as the cell size of a certain tissue is the same in different species, and the organ's volume depends on the number of cells, not their dimensions [10].

Assuming that the endoplasmic reticulum and intracellular organelles have a surface that is tenfold larger than that of the cytoplasmic membrane, we can estimate the ratio of areas of the outer and intracellular membranes to functional receptors' binding sites as (2–3)·10^6:1.

Soluble proteins such as drug acceptors in the cytoplasm, which are superior by volume among all body liquids, and blood plasma occupy totally a much larger surface than the cell membranes[1]. Albumin, having the average plasma concentration of 6·10^{-7} M, accounts for approximately 50% of blood proteins. If an albumin molecule has a spherical shape, then the total surface of this protein in 1 mL of plasma volume can be estimated as ~3·10^{19} nm^2. It can be accepted that the surface of all proteins in plasma takes twice as much as this value. Thus, the ratio of the surface for blood proteins to the surface of binding sites in a specific type of receptors is ~1.3·10^7: 1.

The concentration of anions and cations in the blood plasma is 0.14–0.15 Eq/L, and sodium and chlorine ions make a contribution of ~90% to total inorganic ions. Therefore, the concentration ratio of inorganic ions to specific receptors of a certain type can be estimated as 3·10^7: 1.

The cytoplasm has similar concentrations of proteins and salts as those of blood plasma, albeit their compositions are different. For instance, proteins, inorganic compounds, and nonproteinaceous organic substances constitute, respectively, 67%, 10%, and 23% of total dissolved compounds in the cytoplasm [10] and 70%, 9%, and 21% in the blood plasma [11]. Potassium ions are major cations in the cytoplasm, comprising 0.1–0.15 Eq/L, while phosphates, sulfates, and organic ions are predominant among anions.

However, the role of inorganic ions as effective drug acceptors is relatively small. Because of the high charge concentration, the ions bind to biomolecules without displacement of the ion's hydrate shells. Therefore, the ions' affinity for drugs or receptors is low.

Thus, based on our calculations, the ratio of nonspecific to specific binding sites of a drug molecule can be estimated with good confidence as 10^7: 1, with the accuracy of one decimal order.

7.3 TWO EXPLANATIONS OF DRUG MOLECULES' "PROTECTION" FROM NONSPECIFIC BINDING

Assume that the latency for a given drug's action is dependent on the time required to achieve the necessary receptor blockade and has the longevity of approximately 1 hour. Based on the derived ratio between nonspecific and specific binding sites, we can conclude that the average lifetime of nonspecific complexes must be shorter than 1 hour/10^7 to 0.4 ms. Therefore, the K_d value is estimated using Equation (4.3) as 400 μM, and the free complex formation energy, ΔG, is −(4–5) kcal/mol at 37°C, which is indeed a correspondingly low affinity for drugs.

The accepted explanation for low affinity that a substance–acceptor system achieves the global minimum of free energy, but a stable complex that is not formed without mutual complementarity is not always obvious. Considering huge amounts of nonspecific acceptors estimated above and conformational flexibility of drug molecules, it is difficult to accept that the lack of steric conditions is a hindrance for the formation of a stable complex in all cases.

Another mechanism is possibly involved in complex formation between substances and nonspecific acceptors. The system "substance–acceptor" does not reach the global thermodynamic minimum as a result of kinetic factors, that is, a complex dissociates before a substance binds to the acceptor's all sterically accessible groups when different groups in molecules interact asynchronously.

These explanations for low affinity are not alternatives: binding can proceed through one of these pathways, depending on the structure of the substance and the acceptor. The second interaction pattern assumes the existence of a mechanism by which drug molecules are protected from strong binding to nonspecific targets. In this chapter, we consider the protective role of the molecular electrostatic potential (MEP).

7.4 ROLE OF MOLECULAR ELECTROSTATIC POTENTIAL IN RECOGNITION

The electrostatic field surrounding a molecule is created by atomic charges. Introduction of a polar substituent into a molecule can cause prominent alterations in the MEP, thus changing the orientation of a ligand molecule relatively to the receptor and affecting their binding (the field effect). On the contrary, the electron density on atoms distant for >2–3 ordinary covalent bonds from the substituent varies by a few percent only (the local effect). Minor changes in the electron density are evidenced from small shifts in the vibrational frequencies (approximately 20–30 cm^{-1}) of some bonds, corresponding to 1%–2% changes in the vibrational constants and electron density [12]. The field effect is clearly manifested as the difference between the first and second dissociation constants for aliphatic dicarboxylic acids and polymethylene diamines.

Table 7.1 shows the pK$_a$ and $\Delta\Delta G$ values to show the mutual effect of carboxylic groups separated at different distances.

The difference between the first and second pK$_a$ values of dicarboxylic acids is attributed to the inhibited dissociation of the second carboxyl group by an already formed carboxylate anion. At the distance between the carboxyl groups of ~10 Å (as in suberic acid, n=6), $\Delta\Delta G$ as an indicator to the effect on MEP still exceeds the average kinetic energy of moving molecules in an aqueous solution at 25°C, which is 3/2 RT = 0.89 kcal/mol [1]. This results in the directed movement of molecules due to the attraction of oppositely charged ions. The distance is given for the linear conformation of the molecule, determined by the repulsion of like charged ions at the ends of the methylene chain.

Effects of the electrostatic field for diamines are illustrated in Table. 7.2.

For diamines, the same pattern occurs: if ion charge centers are separated at a distance shorter than 11 Å, then the energy of their interaction is greater than the average kinetic energy of molecules in an aqueous solution.

TABLE 7.1
Dissociation Constants (pK$_{a1}$ and pK$_{a2}$) of Dicarboxylic Acids
HOOC(CH$_2$)$_n$COOH (Water, 25°C) [13]

n	Acid Name	pKa$_1$	pKa$_2$	$\Delta\Delta G$, kcal/mol
0	Oxalic	1.23	4.19	4.03
1	Malonic	2.74	5.33	3.52
2	Succinic	4.16	5.60	1.96
3	Glutaric	4.34	5.28	1.16
4	Adipic	4.43	5.28	1.16
5	Pimelic	4.48	5.32	1.14
6	Suberic	4.51	5.33	1.12

MEP at a given point in the vicinity of a molecule is the Coulomb force that acts on a test proton given the electrostatic energy density at the point, and positively charged regions have a positive sign of the potential. The MEP distribution on a cross-sectional plane is easily interpreted on two-dimensional or three-dimensional graphs with equipotential lines. We previously selected three-dimensional MEP imaging for better visualization in [15]. The MEP calculations and imaging are performed for serotonin, 2-acetoxycyclopropylamine (ACP), adrenaline, gamma-aminobutyric acid (GABA), and 2,4-diaminobutyric acid in the ionized form and diazepam with no charged groups (Figure 7.1).

MEP plots shown in Figures 7.2–7.4 were constructed by MOLDEN [16] using the B3LYP/aug-cc-pvTZ method calculations with the Psi4 package [17].

Calculated MEP values around a molecule are represented graphically in planes parallel to the so-called *baseline plane* defined by some three points of the molecule.

The baseline plane of ACP passes through the nitrogen and two oxygen atoms, constituting the pharmacophore of muscarinic acetylcholine receptor ligands. The baseline plane covers an indole

TABLE 7.2
Dissociation Constants (pK_{a1} and pK_{a2}) of Diamines $H_2N(CH_2)_nNH_2$ (Water, 20°C) [14]

n	pK_{a1}	pK_{a2}	Interatomic Distance (d) N-N, Å	$\Delta\Delta G$, kcal/mol
2	9.97	6.97	3.7	4.0
3	10.65	8.58	5.0	2.8
4	10.84	9.3	6.3	2.1
5	11.05	9.74	7.6	1.8
6	10.93	9.83	8.9	1.5
8	10.99	10.1	11.4	1.2

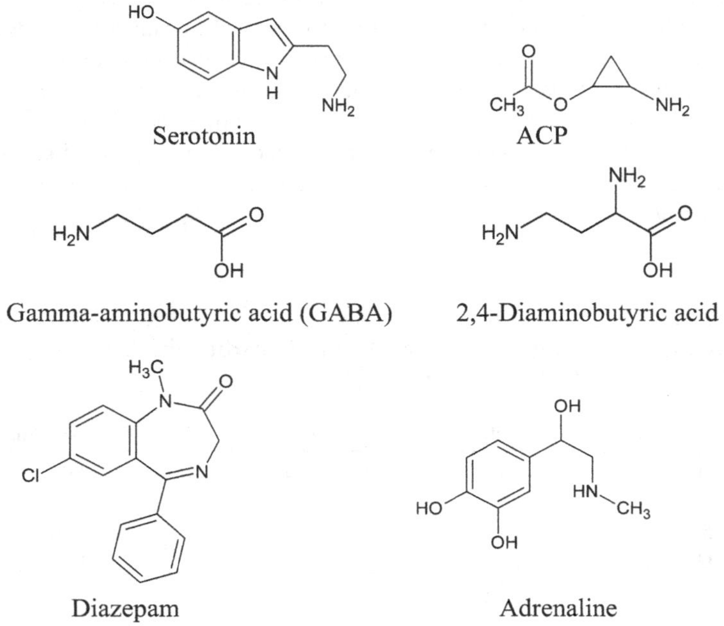

Serotonin   ACP

Gamma-aminobutyric acid (GABA)   2,4-Diaminobutyric acid

Diazepam   Adrenaline

FIGURE 7.1 Chemical structure of molecules for MEP representation.

fragment and a nitrogen atom in the amino group of serotonin, dihydroxyphenyl and amine nitrogen atom in adrenaline, charged centers of terminal anionic and cationic groups, and a carbon chain backbone in GABA and 2,4-diaminobutyric acid molecules. As for the latter molecule, the protonated amino group at position 2 is located out of the plane. In diazepam, the baseline plane comprises atoms with significant partial charges such as imine nitrogen, oxygen, and chlorine.

Based on our calculations that, if a ligand has a charged group, it exerts a major effect on the MEP manifested as a pronounced peak around the charged group area on the electrostatic field (Figure 7.2). The entire molecule is positioned in the region of the positive MEP, while in ligands without a charged group (e.g., diazepam), both positive and negative peaks from the ligand's polar groups are observed (Figure 7.3a). The longer the distance of the MEP image plane from the molecule baseline plane, the fewer the structural features of noncharged polar functional groups displayed as MEP local peaks. Serotonin, adrenaline, and other small monocationic molecules have no notable differences in the MEPs at distances longer than 4 Å from the charge center.

In modeling interaction of monocationic ligands such as serotonin, ACP, adrenaline, and diazepam (Figure 7.1) with a receptor binding site, we used a simplified model with an anionic group only because it has the strongest effect on the receptor's MEP at the initial (distant) interaction stages. Acetate anion mimics the anionic group[2] [18]. A distance of 4.5 Å between the charge centers of the $-NH_3^+$ group in a monocationic ligand and $-COO^-$ in the receptor exceeds the sum of their van der Waals radii in place of the maximum approach between charges (~3.3 Å). As a molecule is approaching to the proximity of a receptor, the positive potential is neutralized, thus leading to the appearance of the MEP that peaks from other uncharged fragments including negatively charged

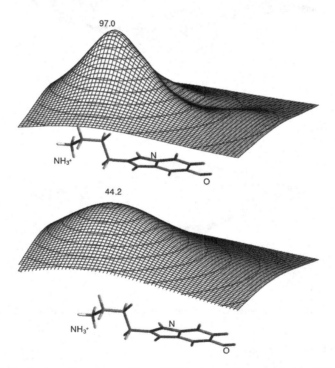

FIGURE 7.2 MEP distribution of serotonin (Figure **7.1**) in the gas phase in planes shifted by 2 Å (a) and 4 Å (b) from the baseline plane.

The MEP values are expressed in kcal/mol (the dielectric constant is not included). The plane dimensions are 20 Å × 20 Å.

parts. This resembles a "molecular stripping" where individual atoms or groups gain a possibility for binding. A search for acceptor's complementary groups is not stochastic and occurs only at a certain distance from the anionic group, that is, a recognition program in the molecular structure starts to occur.

A distinct pattern occurs when a neutral diazepam molecule approaches to the charged receptor active center (Figure 7.3b). Because of the uncompensated receptor's MEP (it remains negative globally), a diazepam molecule is in the field of the same sign as that of the receptor, thus preventing binding.

In general, the MEP of a molecule is uncompensated when the contributions from atoms of uncharged groups do not appear at short intermolecular distances comparable to the van der Waals radius of the carbon atom.

A similar situation occurs when GABA and 2,4-diaminobutyric acid interact with the GABA receptor (Figure 7.4). Approaching of the GABA molecule toward the receptor is followed by MEP compensation, which does not occur in the case of 2,4-diaminobutyric acid, where MEP is uncompensated with a positive peak because of the excessively charged ammonia group. When the interacting partners are in the electrostatic field of the same sign, this situation is unfavorable for binding.

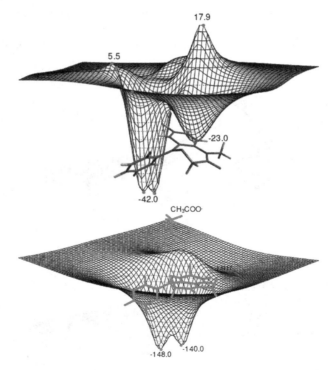

FIGURE 7.3 Modeled interaction between diazepam (Figure **7.1**) and receptors, with one anionic group in the binding site.

(a) MEP at 2 Å from the baseline plane.

(b) MEP with the receptor. The distance between the charge center of the acetate anion and the baseline plane is 4.5 Å. MEP is calculated at 2 Å from the baseline plane.

The MEP values are expressed in kcal/mol (the dielectric constant is not included). The plane dimensions are 20 Å × 20 Å.

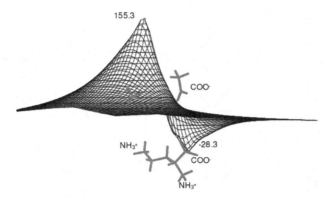

FIGURE 7.4 Modeling of the 2,4-diaminobutyric acid (Figure **7.1**) interaction with the GABA receptor.

The distance from the baseline plane to the ammonia and acetate ions is 4.5 Å. MEP is calculated for a 2 Å distance closer to the receptor-mimicking groups from the baseline plane. The centers of opposite charges of the substance and receptor's mimicking ion groups are on the same line perpendicular to the baseline plane, except for an excess $-NH_3$ group in the second position.

The MEP values are expressed in kcal/mol (the dielectric constant is not included). The plane dimensions are 20 Å × 20 Å.

During the interaction of diazepam with its receptors containing a charged group in the active site as well as in the case of 2,4-diaminobutyrate and the GABA receptor, the interacting partners appear to be in the electrostatic field of the same sign, and this situation is unfavorable for binding. Indeed, the affinity of 2,4-diaminobutyric acid for the GABA receptor drops by more than 10^4 times as compared to GABA itself. The uncompensated MEP impedes or prevents complex formation in many cases. As seen in Figures 7.2 and 7.3, this is the case when the number of charged groups in a substance and receptor's binding site is not the same.

In addition to three-dimensional MEP representation, there are many others such as contour color map or molecular surface color map. Concise and accurate representation of MEP topography as a set of critical points (CPs), such as saddle, maxima, and minima, was proposed in [19] (see also a recent review in [20]). Molecular recognition was explained in terms of changes in CP configuration when molecules approach each other. Pyridine and benzene dimers, as well as DNA base pair interactions, were considered. Such changes in CPs occur at distances smaller than 4–5 Å, in accordance with our estimates. The model presented in [19] resembles our model of the MEP role in the recognition process.

7.5 MOLECULAR ELECTROSTATIC POTENTIAL AND EXPERIMENTAL AFFINITY

How do the conclusions in the previous subsection match to affinities of compounds for different receptors?

Widely varying values of affinities for different receptors are shown in Table 7.3. Based on the specific ligand structure, we can classify receptors into the following types, depending on the charge sign and the number of charged groups in the receptor binding site:

I. Receptors with one anionic group (muscarinic acetylcholine receptors, dopamine, adrenaline, and serotonin receptors).
II. Receptors with two anionic groups (nicotinic acetylcholine receptor and receptors for bisonium compounds).

TABLE 7.3
Substances' Affinity Profiles for Different Receptors

Compound	K_d, M									
	mAChR	D_1	D_2	α_1	α_2	H_1	5-HT_1	5-HT_2	GABA-R	*
Promethazine	$2.1 \cdot 10^{-8}$ [22]	$1.8 \cdot 10^{-6}$ [21]	$2.4 \cdot 10^{-7}$ [21]	$5.5 \cdot 10^{-8}$ [23]	$1.2 \cdot 10^{-6}$ [24]	$2.9 \cdot 10^{-9}$ [25]				
Promazine	$9.6 \cdot 10^{-8}$ [26]	$7.9 \cdot 10^{-7}$ [27]	$7.2 \cdot 10^{-8}$ [28]	$1.1 \cdot 10^{-8}$ [29]	$9.0 \cdot 10^{-7}$ [30]	$3.0 \cdot 10^{-8}$ [25]	$3.1 \cdot 10^{-6}$ [31]	$1.6 \cdot 10^{-7}$ [32]		
Diphenhydramine	$1.2 \cdot 10^{-7}$ [21]	$>1 \cdot 10^{-5}$ [27]	$1 \cdot 10^{-5}$ [21]	$3.5 \cdot 10^{-6}$ [23]	$3.5 \cdot 10^{-6}$ [24]	$1.7 \cdot 10^{-8}$ [25]				
Cyproheptadine	$1.9 \cdot 10^{-8}$ [33]	$1.1 \cdot 10^{-7}$ [27]	$3.1 \cdot 10^{-8}$ [33]	$1.0 \cdot 10^{-7}$ [33]	$7.6 \cdot 10^{-7}$ [33]	$2.7 \cdot 10^{-9}$ [33]	$7.0 \cdot 10^{-7}$ [33]	$6.5 \cdot 10^{-9}$ [33]	$>1 \cdot 10^{-6}$ [33]	
Morphine	$>1 \cdot 10^{-6}$ [33]		$>1 \cdot 10^{-6}$ [33]	$>1 \cdot 10^{-6}$ [33]	$>1 \cdot 10^{-6}$ [33]	$>1 \cdot 10^{-6}$ [33]	$>1 \cdot 10^{-6}$ [33]	$>1 \cdot 10^{-6}$ [33]	$>1 \cdot 10^{-6}$ [33]	$6.5 \cdot 10^{-9}$ [33]
Atropine	$5.5 \cdot 10^{-10}$ [34]	$>1 \cdot 10^{-5}$ [27]	$1.0 \cdot 10^{-5}$ [35]	$6.6 \cdot 10^{-7}$ [23]		$2.0 \cdot 10^{-6}$ [25]		$4.3 \cdot 10^{-6}$ [36]		
Hexamethonium	$>1 \cdot 10^{-4}$ [37]									$8.2 \cdot 10^{-9}$ [38]
GABA	$>>1 \cdot 10^{-5}$ [39]	$>1 \cdot 10^{-4}$ [40]	$>1 \cdot 10^{-4}$ [41]	$>1 \cdot 10^{-5}$ [23]	$>1 \cdot 10^{-4}$ [42]	$>4.8 \cdot 10^{-5}$ [43]	$>1 \cdot 10^{-4}$ [44]	$>1 \cdot 10^{-3}$ [45]	$1.6 \cdot 10^{-8}$ [46]	
2,4-Diaminobutyric acid									$>1 \cdot 10^{-4}$ [46]	
Glycine	$>>1 \cdot 10^{-5}$ [39]	$>1 \cdot 10^{-5}$ [47]			$>1 \cdot 10^{-4}$ [42]				$>1 \cdot 10^{-4}$ [46]	
Glutamine acid	$>>1 \cdot 10^{-5}$ [39]	$>1 \cdot 10^{-4}$ [40]	$>1 \cdot 10^{-4}$ [37]	$>1 \cdot 10^{-4}$ [23]				$>1 \cdot 10^{-3}$ [45]	$>1 \cdot 10^{-4}$ [46]	$5 \cdot 10^{-7}$ [48]
Diazepam	$>>1 \cdot 10^{-5}$ [39]	$>1 \cdot 10^{-4}$ [40]	$>1 \cdot 10^{-5}$ [47]		$>1 \cdot 10^{-4}$ [42]		$>1 \cdot 10^{-5}$ [41]	$9.1 \cdot 10^{-6}$ [45]		$2 \cdot 10^{-9}$ [49]
Chlordiazepoxide	$>>1 \cdot 10^{-5}$ [39]	$>1 \cdot 10^{-5}$ [47]	$>1 \cdot 10^{-5}$ [47]		$>1 \cdot 10^{-4}$ [42]					
Estradiol			$>1 \cdot 10^{-3}$ [50]	$1.8 \cdot 10^{-4}$ [50]						$2 \cdot 10^{-10}$ [51]
Tetrahydrocannabinol	$>>1 \cdot 10^{-5}$ [39]	$>>1 \cdot 10^{-5}$ [47]	$>1 \cdot 10^{-5}$ [47]							$2.4 \cdot 10^{-8}$ [52]

mAChR, muscarinic cholinergic receptor; D_1 and D_2, subtypes of dopamine receptors; α_1 and α_2, subtypes of adrenaline receptors; H_1, subtype of histamine receptors; 5-HT_1 and 5-HT_2, subtypes of 5-hydroxytryptamine (serotonin) receptors; GABA-R, receptors of gamma-aminobutyric acid; *, specific receptor or binding site of the ligand (for diazepam, it is the benzodiazepine allosteric binding site on the GABA receptor).

III. Receptors with one anionic group and one cationic group (GABA and glycine receptors).
IV. Receptors with one anionic group and two cationic groups (glutamine acid receptors).
V. Receptors without charged groups in the specific binding site (receptors for benzodiazepine and steroid hormones)[3].

According to the proposed classification, cyproheptadine (Figure 5.3), promethazine, promazine, diphenhydramine, atropine, and morphine (Figure 7.5) are ligands for type I receptors; hexamethonium (Figure 7.5) for type II; GABA (Figure 7.1) and glycine (Figure 7.5) for type III; glutamine acid (Figure 7.5) for type IV; and diazepam (Figure 7.1), chlordiazepoxide, estradiol, and tetrahydrocannabinol (Figure 7.5) for type V receptors. The fact that we determined the number of

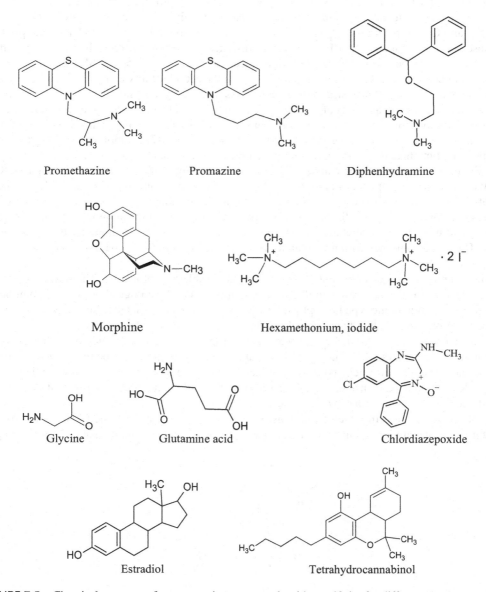

FIGURE 7.5 Chemical structure of representative compounds with specificity for different receptors.

charged groups in the active center of the receptors indirectly does not affect the accuracy of the determination, as the removal or addition of charged groups in the ligand molecules drastically reduces their affinity for this receptor.

Noteworthy, compounds have low affinity for nonspecific receptors and can have either low or highly affinity to specific receptors of their "native" type. This conclusion that follows from the affinity range for the considered compounds agrees with the conclusions made from the role of MEP in molecular recognition.

GABA, diazepam (Figure 7.1), tetrahydrocannabinol, glycine, glutamic acid, morphine (Figure 7.5), and some other compounds possess high affinity for their specific receptors only. On the contrary, promethazine, promazine (both depicted in Figure 7.5), and cyproheptadine (Figure 5.3) notably have affinity (K_d 10^{-8}–10^{-7} M) for muscarinic cholinergic receptors, dopamine, adrenaline, histamine, and serotonin receptors belonging to the same receptor type. However, cyproheptadine does not display high affinity for nonspecific receptors; its values are low for melatonin ($K_d > 1 \cdot 10^{-4}$ M) [53] and GABA receptors ($K_d > 1 \cdot 10^{-6}$ M). Glycine is not bound to the GABA receptor possibly because of the uncompensated MEP because the distance is different between charged groups (in the same amounts) in the glycine molecule and the receptor. A value of $K_d > 10^{-6}$ M in experiments is a common indicative for a low affinity ligand, although the actual affinity can be substantially lower.

Some known drugs have structural peculiarities that bring a contradiction to the proposed statement that the same amounts of charged groups in the ligand and receptor are crucial for the formation of a stable complex. Thus, trifluoperazine (Figure 7.6), as a diamine with two basic nitrogen atoms in a piperazine moiety, exerts high affinity ($K_d = 2.1$ nM) [47] for monoanionic D_2-dopamine receptors that mediate the neuroleptic (antipsychotic) effect of this ligand.

This supposed contradiction can be explained by the fact that trifluoperazine ($pK_{a1} = 8.40$; $pK_{a2} = 3.90$) exists almost completely in the monocationic form in an organism (pH = 7.3) [54]. As shown in Table 7.2, an effect of the electrostatic field that impedes protonation of a secondary amino group is weakened by enlarging the distance between two amino groups. If the amino groups are separated by four and more methylene chains ($n \geq 4$), in water, diamines are predominantly in the dicationic form. Therefore, diamines with $n > 3$ cannot be bound effectively to monoanionic receptors (type I receptor).

The drug molecules usually contain several atoms capable of protonating. The formation of several charged centers would result in the disappearance of the physiological activity of compounds selective for monoanionic type I receptors. However, drugs are constructed in such a way that not all atoms are protonated under physiological conditions (see Chapter 1 also).

Consider the antiarrhythmic drug procainamide (Figure 7.7). Its molecule contains four atoms with lone electron pairs—three nitrogen atoms and one oxygen atom may be protonated. Of these, the aniline and amide nitrogen and the oxygen atom are in conjugation, which causes redistribution of the electron density. Consequently, the basicity of the aniline and amide nitrogen atoms drastically decreases, and their pKa ≈ 2 (see Table 1.2).

Despite the increase in the electron density on the oxygen atom, its basicity is low because of greater electronegativity and nuclear charge that contribute to weak donor capacity of the lone pair

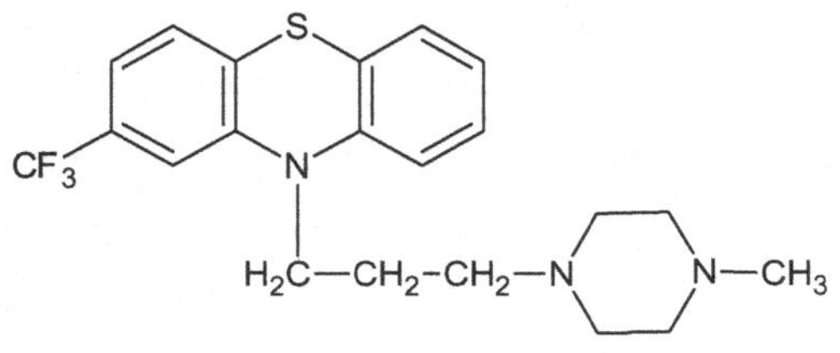

FIGURE 7.6 Chemical structure of trifluoperazine.

when compared with the nitrogen atom. The nitrogen atom of the alkyl amino group is not involved in conjugation and has a high basicity (pKa = 9.2 [55]), which is enhanced by the inductive effect of the linking methyl groups.

Thus, the aniline and amide nitrogen atoms are not protonated under physiological conditions (pH=7.3), compared with the nitrogen atom of the amino group. The mesomeric effect can be called "internal" because this is due to the mobility of the electrons in the covalent bonds, in contrast to the "external" field effect of the receptor protein groups.

Molecules of the α-amino acids glycine and glutamic acid (Figure 7.5), containing charged groups in the side chain, are completely ionized in aqueous solutions. Therefore, a stable complex containing monoanionic type I receptors is not formed because of the different numbers of equally signed charged groups in these molecules and the receptor's binding site.

Hexoprenaline, a tocolytic agent with the commercial name Gynipral (Figure 7.8), has two amino groups, although the binding site of the targeted β$_2$-adrenaline receptors contains one anionic group as can be judged from their specific typical monocationic ligands such as orciprenaline, fenoterol, terbutaline, and others.

Based on the bis-symmetric structure of hexoprenaline, it may be assumed that the anionic groups of the binding sites are positioned at distances of 8–10 Å and belong to neighboring β$_2$ adrenergic receptors in some tissues at least.

Noteworthy, the nicotinic acetylcholine receptor, containing two anionic groups in its functional site, displays remarkable affinity for dicationic molecules (e.g., tubocurarine and hexamethonium) but also binds to monocationic ligands such as acetylcholine and nicotine. In this case, a large distance between anionic groups (14 Å and more) [56, 57] is important. Because of such a structure of the functional site, small monocationic molecules can bind to one of the anionic centers, although the formed complex has a weakened strength. Low affinity of acetylcholine in this case corresponds to short complex lifetimes (see Chapter 6.1) and favors fast receptor reactivation.

We have described the mechanisms that protect the charged drugs from unproductive binding. However, among the drugs, there are many neutral substances acting on specific receptors. Are there any peculiarities in recognition for such drugs and what role does the MEP play? The algorithm for finding its "own" specific receptor seems the same as that of charged drug molecules, although the group or atom with the most pronounced partial atomic δ-charges performs the role of the charged group. The presence of dipole polar groups in the structure of neutral drugs appears mandatory. The principle of asynchronism (stepwise process) is the basis for the recognition algorithm for uncharged drug molecules. The primary interaction is conducted by the

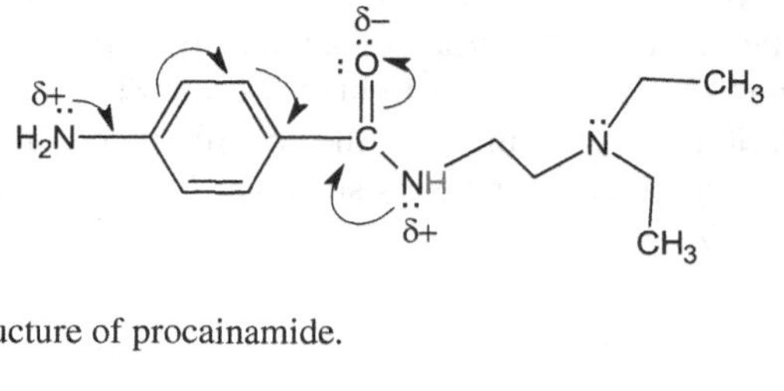

FIGURE 7.7 Chemical structure of procainamide.

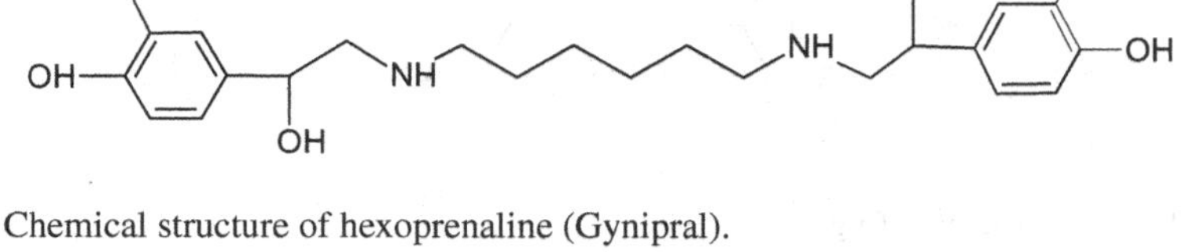

FIGURE 7.8 Chemical structure of hexoprenaline (Gynipral).

dipole groups with the most pronounced partial atomic charges, after which other dipole groups and finally the "anchor" nonpolar groups begin to function. Moreover, it is possible that the dipole moment of the entire molecule plays an important role in the recognition of neutral drug molecules. For example, consider the distribution of charges in neutral diazepam **7.5**. As a result of the mesomeric effect (see Chapter 1), the most pronounced δ^--charge in the molecule is located on the carbonyl oxygen atom. The opposite δ^+-charge is distributed on atoms participating in the mesomeric effect. The chlorine atom also bears a pronounced δ^--charge. The "anchor" phenyl groups would bind to a specific receptor at contact van der Waals distances, that is, at the end of the binding process.

7.6 STAGES WHEN A MOLECULE IS ORIENTED AND ATTACHED TO A RECEPTOR: A CASE OF mAChR LIGANDS

Specific interactions during N-methyl-4-piperidyl ester of di-thienyl-glycolic acid C-**4.11** (Figure 7.9) binding to muscarinic cholinergic receptors (mAChR) are illustrated in Figure 7.10. In this ligand molecule, the following fragment can be discerned by the type of interaction with mAChR: (*i*) a charge head with a protonated nitrogen atom and linked aliphatic groups; (*ii*) polar C=O, C-O, and OH groups; and (*iii*) a nonpolar moiety consisting of two thienyl groups. To consider the fragment contribution at different stages during binding to the specific receptor, we calculated the interaction energy of the ligand C-**4.11** chemical groups with the receptor protein at a range of equilibrium distances. The energy of ion–ion, dipole–dipole, and dispersion interactions is dependent on the distance as $1/d$, $1/d^3$, and $1/d^6$, respectively, and these interactions appear to participate in a stepwise manner. Electrostatic energy is calculated using Equation (1.3), with the local dielectric constant being estimated using Equation (1.35) (see details in Chapter 1).

The cationic group begins to get attracted to the receptor's anionic active site from the most distant point. Electrostatic attraction energy begins to exceed the heat motion energy (horizontal line shown in Figure 7.10) at a distance of 6–7 Å between the ligand and receptor van der Waals surfaces. Up to a distance of ~2 Å between the van der Waals surfaces, the attraction energy of the cationic group surpasses the attraction energy of all the remaining molecular groups. At a distance of 2 Å, the ligand's and receptor's MEPs are compensated considerably, thus favoring the binding of uncharged polar groups. Starting from a distance of ~2 Å and below between the van der Waals surfaces, an orienting effect of polar groups appears as a result of dipole–dipole interactions. In the last turn, thienyl groups with the most anchoring capability start interacting with the receptor at near-equilibrium distances, and a ligand–receptor complex is abruptly strengthened. Notably, a sequence of joining the interaction: "cationic group–dipole group–nonpolar group" is not the same as the order of anchoring extent: "nonpolar group–cationic group–dipole group." Figure 7.19 sheds some light to the interaction energy of different ligand groups at stages of orientation and placement on the specific receptor. The former stage corresponds to the distance between the van der Waals surfaces $d_{vdW} > 0$. The placement stage occurs at the closest contact ($d_{vdW} = 0$).

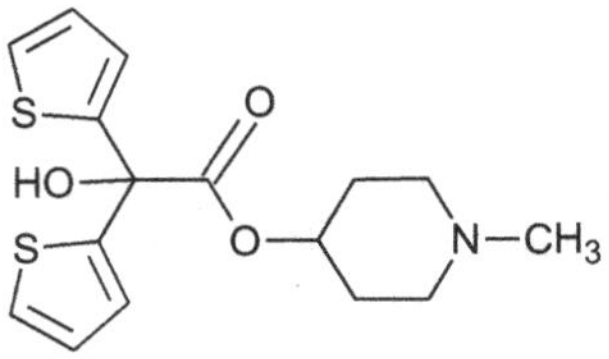

FIGURE 7.9 Chemical structure of ligand C-4.11.

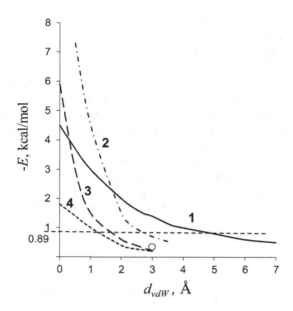

FIGURE 7.10 Modeling dependence of the attraction energy of the ligand C-**4.11** fragments to mAChR on the distance d_{vdW} between van der Waals surfaces of the interacting ligand and the receptor (d_{vdW}=0 means surfaces adjoining). Lines correspond to (**1**) the cationic group; (**2**) all fragments without the cationic group; (**3**) thienyl groups; and (**4**) the carbonyl group.

7.7 RECOGNITION AS A METHOD OF TRIALS AND ERRORS

Terminologically, the recognition refers to an action by the method of trials and errors. Being introduced from the field of human relations into molecular pharmacology, this term is the most felicitous, pertaining to not only a result but also a process, when a molecule seeks its specific receptor among a great variety of nonspecific acceptors.

An analogy to the recognition of previously an unknown person in the crowd is apparent. Assume that you need to meet a person with a height of approximately 6 feet in a gray suit, with a bag and a newspaper in his hand when cell phones were not used. First, you confine the search by the height of people. If the other features, insignificant before (such as suit color, bag, or newspaper), will coincide you can think that the recognition is OK. Likewise, a molecule runs an algorithm of its behavior and selects first the basic feature of a specific receptor—a certain number of complementary ion groups. Then, mutual neutralization of both ligand and receptor MEPs creates the conditions to look for details of the receptor structure—the position of partial atomic charges in polar groups on the binding site and its topography, which have been insignificant at the first recognition stage but become crucial at the next stages. We have already considered the reasons for why drugs do not bind to their nonspecific receptors. The compensated MEP of a substance and a receptor during the specific interaction creates the conditions for subsequent complementary binding stages, culminating with the formation of a stable complex. To correct recognition errors, partial contributions of "anchor" (the most strongly interacting) aromatic groups should not be too high at distant stages of interaction with a receptor.

7.7.1 CHARGED GROUP RECOGNITION

When a cationic molecule enters the area of the receptor's anionic active site and begins getting attracted, it begins moving directly and implements a recognition program encoded in its structure.

At the stage of distant interaction, the behavior of this molecule is determined by a charged group. Although ion–ion interactions are not specific (i.e., any anionic molecule or residue attracts the cationic molecule), the MEP of the cation prevents the molecule from binding to various acceptors lacking the anionic group. When a cationic group associates with the protein's anionic group, a molecule searches only complementary sites at a definite distance from the anionic active center. This search is not stochastic. Strengthening or dissociating a primary complex depends on whether or not the binding of the charged group is reinforced by the attraction of other groups. For the correction of errors in ligand–receptor interactions, the energy of the charged group binding must be relatively low in magnitude, despite its long-range action.

The data presented in Table 7.4 show partial contribution from the charged group to the free energy of the ligand C-**4.11** binding to mAChR, as compared with other examples of ligand–protein binding. Indeed, the partial contribution of a charged group to the binding free energy is no more than ~1.5–2.0 kcal/mol. According to relationship in Equation (6.3), a ΔG value of -2 kcal/mol corresponds to the lifetime of ~10^{-6} s of a complex. Hence, the total residence time of a substance on

TABLE 7.4
Partial Contribution of the Charged Group into the Binding Free Energy

Compound		Protein	K_d (K_i), M	K^0_d/K^+_d	$-\Delta\Delta G$, kcal/mol	Ref.
C-4.11			$5.6 \cdot 10^{-11}$			
		mAChR		10	1.5	[58]
C-4.11 uncharged analog			$5.8 \cdot 10^{-10}$			
$(CH_3)_2 N^+HCH_2CH_2OCOCH_3$ $(CH_3)_2 CHCH_2CH_2OCOCH_3$		AChE	$1 \cdot 10^{-3}$ $8 \cdot 10^{-3}$	8	1.28	[59]
$(CH_3)_2 N^+HCH_2CH_2OH$ $(CH_3)_2 CHCH_2CH_2OH$		AChE	$4.5 \cdot 10^{-4}$ $1.4 \cdot 10^{-2}$	31	2.12	[59]
	Cation, pH 6		$1.8 \cdot 10^{-8}$			
	———	AChE		25	1.99	[60]
	Neutral base, pH 10		$4.5 \cdot 10^{-7}$			
Eserine						
		Antibody	—	8	1.15	[61, 62]

nonspecific acceptors during an interaction between only two charged groups is small, and this is sufficient to correct for errors during recognition.

This confirms that recognition mechanisms governed by a structure of small molecules are versatile and simple. The simplicity principle helps a drug molecule to find an efficient solution to a difficult problem, such as specific receptor selection.

7.7.2 RECOGNITION OF DIPOLE (POLAR) GROUPS

Polar diatomic groups (e.g., carbonyl) or groups from more than two atoms or functional groups can be referred as having the prominent dipole moments. Typically, a dipole group contains a heteroatom, or a carbon–carbon bond formed by sp^2- or sp-hybridized carbon atoms. After charged groups, dipole groups are the next participants in the long-range action, although their effective radius of interaction is smaller than that of ions.

In contrast to a charged group, even small violations of the optimal orientation cause a drastic reduction in dipole attraction or even repulsion (see Chapter 1). Dipole groups spot the absence or presence of complementary sites located at a certain distance from the receptor's anionic active center. It would be worthwhile to verify that physiologically active compounds, acting on various receptors, differ in the position of dipoles relatively to the cationic group.

One can find statements in the relevant literature that a dipole group in a drug molecule is not significant because, for instance, the substitution of oxygen for methylene or methyl results in a slight decrease in the affinity to a specific receptor in several cases. However, it should be kept in mind that the selectivity of substances is greatly reduced because of such molecular modifications.

At equilibrium distances, the dipole attraction energy is large enough to keep them together. For example, the potential energy of attraction between a properly oriented carbonyl groups can reach 2 kcal/mol due to the small local dielectric constant of the medium in the vicinity of dipole groups. At equilibrium distances, the dielectric constant can be assumed equal to 1 (see Chapter 1). Polar groups have no strong orienting effect on water molecules; therefore, dielectric screening is small. Dipoles of polar groups play various roles in distant and contact interactions:

- Electrostatic interaction between dipoles is important during ligand orientation near a receptor,
- Polar groups can form hydrogen bonds with the receptors at the stage of ligand placing within the receptor's binding pocket.

Thus, during interaction with the specific receptor, polar groups with a substantial dipole moment orient and attach to a ligand molecule on the receptor surface in the position that permits a complementary binding of all groups of the ligand.

As with compound C-**4.11**, dipoles of carbonyl and ether groups with the charged amino group located at a certain distance from each other constitutes the pharmacophore (see details in Chapter 9), the main structural element for the recognition of the muscarinic cholinergic receptor. The hydroxyl group contributes -1.5 kcal/mol to the binding free energy (see Table 6.6 in Chapter 6), and this value can be attributed to hydrogen bond formation. However, there are mAChR antagonists with much smaller partial contribution of the hydroxyl group, and for them, different binding mechanisms might exist (see details in Chapter 6), but the recognition process is the same.

7.7.3 RECOGNITION OF BULKY NONPOLAR GROUPS

Molecular events during the recognition (which we may call a recognition scenario) of the specific receptor by the ligand C-4.11 (Figure 7.9) are ordered so that most anchoring groups are bound to a receptor protein in the last turn; otherwise, it would be difficult to eliminate errors during recognition. This order is attributed to the fact that dispersion interactions come to play at equilibrium distances only. Binding of bulky nonpolar groups present distant from the charged groups (i.e., thienyl groups in C-**4.11**) is the final event when searching for a specific receptor. As nonpolar groups are bound to the receptor during direct contact, its interaction energy is strongly dependent on the conformation and mutual orientation of the participating groups.

Long-range charged and polar groups play the most active role in finding the receptor, whereas nonpolar groups passively participate in this process and mainly define the affinity (K_d) of compounds. Being at an unfavorable conformational state and having extra optimal size, nonpolar groups can inhibit the binding of any other molecular groups to the specific receptor. The passive participation of nonpolar anchoring groups in molecular recognition can be supported by high-selectivity of mAChR agonists such muscarine, acetylcholine, arecoline, 5-methylfurmethide, and others, which have none of these groups. At the same time, nonpolar and slightly polar groups are very important to form the pharmacophore structure because of the optimization of distances between key charges (see further discussion in Chapter 9).

7.8 CONCLUSION

A process of drug–acceptor binding can be divided into three stages:

1) A long-distant interaction, without establishing contact between molecules.
2) An interaction under incomplete contact when a molecule is oriented on the acceptor's surface.
3) Complex formation when the molecule is set on the acceptor's surface or within the pocket.

At some distance, a molecule starts moving directly relative to the surface of a protein or another acceptor due to intermolecular forces. As already noted, this interaction becomes efficient if its energy exceeds the kinetic energy of a molecule. Distantly acting charged groups in water molecules begin attracting or repulsing at distances of 10–11 Å between their charge centers, that is, when monomolecular hydrate shells are still present.

It can be accepted with good accuracy that a distant (long-range) interaction between two molecules is solely electrostatic, and it can be expressed through an interaction of their electrostatic potentials (fields). We may say that a ligand and a receptor "see" each other from afar through the MEP. Superposition of electrostatic potentials with different signs orients a molecule relatively to an acceptor even at the first stage so that the mutual approach between charged groups occurs first. The initial stages of approaching are not affected by other molecular fragments located in the positive potential region within an efficient radius of ~10 Å from the cationic group charge center in an aqueous solution (comparison of data on diamines and dicarboxylic acids shown in Tables 7.2 and 7.3). This region is favorable for a stronger organization of water molecules around hydrophobic groups. Such an increase in "hydrophilicity" declines the probability of binding of the latter groups to arbitrary hydrophobic sites of proteins.

Owing to MEP, most nonspecific acceptors are recognized by a molecule at the primary interaction stages, including the stage of a long-range interaction.

In general, the stage of ligand approaching and orientation on a protein is thermodynamically unfavorable because of entropic losses caused by inhibition of translational and partially rotational degrees of freedom. However, the specific MEP of a protein's active site, including the preorganization of a protein's electrostatic environment in enzymes [63, 64], may favor intermolecular forces acting at distances far from intermolecular contact between substances having charged groups.

Hence, the major role of MEP in molecular recognition is reduced to shortening the lifetime of a "compound–nonspecific acceptor" complex. It is a unique feature that MEP makes drug molecules that can distinguish between acceptors at distances far from intermolecular contact. This MEP-mediated mechanism is likely one of the major tools to correct errors during recognition. Recognition of nonspecific acceptors based on MEP seems to be simpler and more powerful than the creation of steric barriers for many thousands of low-molecular endogenous and exogenous compounds including drugs.

The described mechanism by which drugs are protected from being lost in an organism may encounter the following objection. There is a certain amount of high-energy "hot" molecules, which is subject to the Maxwell statistical distribution. Such molecules overcome barriers from the unfavorable MEP and dehydration. These views are taken from the field of gas-phase reactions. However, the aquatic environment significantly reduces the energy scale of this distribution.

Another misconception exists. For the question "What will happen if the charged part of a drug molecule consists of not one but two ion groups of the same charge sign?", we usually get the answer that the affinity and selectivity for this receptor will increase. In fact, the affinity and selectivity will disappear.

NOTES

1 Thus, error in the determination of membrane surface of the endoplasmic reticulum has not too much importance.
2 Anionic group of the orthosteric active site (binding pocket) of many G-protein–coupled receptors (GPCRs) is the **Asp** residue side chain.
3 In addition to the listed receptor classes, there are other classes such as prostaglandin receptors with one cationic group in the binding site.

REFERENCES

[1] D. Metzler, Biochemistry: The Chemical Reactions of Living Cells, vol. 1, Amsterdam: Elsevier, 1980, p. 332.
[2] F. Dukhovich, E. Gorbatova and V. Kurochkin, *Pharm. Chem. J.*, vol. 35, p. 260, 2001.
[3] A. Kiselev, A. Iogansen and K. Sakodynskii, Physico-Chemical Applications of Gas Chromatography, Moscow: Khimiya, 1973, p. 255 (in Russian).
[4] M. Mashkovskii, Lekarstvennye sredstva (Drugs), vol. 1, Moscow: Novaya Volna, 2000, p. 140 (in Russian).
[5] F. Huho, Neurochemistry (Russian transl.), Moscow: Mir, 1990, p. 249.
[6] F. Dukhovich, E. Gorbatova, V. Kurochkin and V. Petrunin, *Russian Chemistry Journal*, vol. 43, no. 5, p. 12, 1999 (in Russian).
[7] R. Bonner, F. Barrantes and T. Jovin, *Nature*, vol. 263, p. 429, 1976.
[8] F. Barrantes, *Biochem. Biophys. Res. Commun.*, vol. 72, p. 479, 1976.
[9] A. Albert, Selective Toxicity, London: Chapman and Hall, 1983.
[10] E. De Robertis, W. Nowinski and F. Saez, Cell Biology, Philadelphia: Saunders, 1970.
[11] A. White, P. Handler and E. Smith, Principles of Biochemistry, vol. 3, New York: McGraw-Hill, 1973, p. 1158.
[12] L. Gribov and S. Mushtakova, Quantum Chemistry, Moscow: Gardariki, 1999, p. 179 (in Russian).
[13] G.N. Freidlin, Aliphatic Dicarboxylic Acids, Moscow: Khimiya, 1978 (in Russian).
[14] P. Kan, "Aliphatic amines," In: *Condensation Monomers*, J. Stille and T. Campbell, Eds., New York: Wiley-Interscience, 1972.
[15] F. Dukhovich and M. Darkhovskii, *J. Mol. Recogn.*, vol. 16, p. 191, 2003.
[16] G. Schaftenaar and J. Noordik, *J. Comp. Aid. Mol. Design*, vol. 14, p. 123, 2000.
[17] R. Parrish, L. Burns, D. Smith, et al., *J. Chem. Theory Comput.*, vol. 13, p. 3185, 2017.
[18] A. Venkatakrishnan, X. Deupi, G. Lebon, C. Tate, G. Schertler and M. Babu, *Nature*, vol. 494, p. 185, 2013.
[19] D. Roy, P. Balanarayan and S. Gadre, *J. Chem. Sci.*, vol. 121, p. 815, 2009.
[20] S. Gadre, C. Suresh and N. Mohan, *Molecules*, vol. 26, p. 3289, 2021.
[21] S. Peroutka and S. Shyder, *Lancet*, vol. 319, no. 8273, p. 658, 1982.
[22] P. Andersen, F. Gronvald and J. Jansen, *Life Sci.*, vol. 37, p. 1971, 1985.
[23] D. U'Prichard, D. Greenberg and S. Snyder, *Mol. Pharmacol.*, vol. 13, p. 454, 1977.

[24] P. Timmermans, J. van Kemenade, Y. Harms, G. Prop, E. Graf and P. Van Zwieten, *Arch. Int. Pharmacodyn. Ther.*, vol. 261, p. 23, 1983.

[25] V. Tran, R. Chang and S. Snyder, *Proc. Nat. Acad. Sci. USA*, vol. 75, p. 6290, 1978.

[26] S.-C. Lin, K. Olson, H. Okazaki and E. Richelson, *J. Neurochem.*, vol. 46, p. 274, 1986.

[27] G. Faedda, N. Kula and R. Baldessarini, *Biochem. Pharmacol.*, vol. 38, p. 473, 1989.

[28] P. Seeman and H. Niznik, *ISI Atlas of Sci. Pharmacol.*, vol. 2, p. 161, 1988.

[29] S. Peroutka, D.U. Prichard, D. Greenberg and S. Snyder, *Neuropharmacol.*, vol. 16, p. 549, 1977.

[30] B. Syoholm, R. Voutilainen, K. Luomala, J.-M. Savola and M. Scheinin, *Eur. J. Pharmacol.*, vol. 215, p. 109, 1992.

[31] S. Mason and G. Reynolds, *Eur. J. Pharmacol.*, vol. 221, p. 397, 1992.

[32] H. Meltzer and J. Hash, *Pharmacol. Rev.*, vol. 43, p. 587, 1991.

[33] J. Leysen, F. Awouters, L. Kennis, P. Laduron, J. Vandenberk and P. Janssen, *Life Sci.*, vol. 28, p. 1015, 1981.

[34] E. Hulme, N. Birdsall, A. Burgen and P. Menta, *Mol. Pharmacol.*, vol. 14, p. 737, 1978.

[35] H. Hall, C. Kohler and L. Gawell, *Eur. J. Pharmacol.*, vol. 111, p. 191, 1985.

[36] R. Lyon, K. Davis and M. Titeler, *Mol. Pharmacol.*, vol. 31, p. 194, 1987.

[37] P. Laduron, M. Verwimp and J. Leysen, *J. Neurochem.*, vol. 32, p. 421, 1979.

[38] V. Skok, A. Selyanko and V. Derkach, Neuronal Acetylcholine Receptors, Moscow: Nauka, 1987, pp. (a) 149; (b) 258 (in Russian).

[39] H. Yamamura and S. Snyder, *Proc. Nat. Acad. Sci. USA, Biol. Sci.*, vol. 71, p. 1725, 1974.

[40] W. Billard, V. Ruperto, G. Crosby, L. Iorio and A. Barnett, *Life Sci.*, vol. 35, p. 1885, 1984.

[41] M. Baudry, M. Martres and J. Schwartz, *Naunyn-Schmiedeberg`s Arch. Pharmacol.*, vol. 308, p. 231, 1979.

[42] P. Ernsberger, M. Meeley, J. Mann and D. Reis, *Eur. J. Pharmacol.*, vol. 134, p. 2, 1987.

[43] J.-M. Palacios, J.-C. Schwartz and M. Garbarg, *Eur. J. Pharmacol.*, vol. 50, p. 443, 1978.

[44] H. Shimizu, T. Tatsuno, A. Hirose, H. Tanaka, Y. Kumasaka and M. Nakamura, *Life Sci.*, vol. 42, p. 2419, 1988.

[45] J. Leysen, C. Niemegeers, J. Van Nueten and P. Laduron, *Mol. Pharmacol.*, vol. 21, p. 301, 1982.

[46] S. Enna and S. Snyder, *Mol. Pharmacol.*, vol. 13, p. 442, 1976.

[47] D. Burt, I. Creese and S. Snyder, *Mol. Pharmacol.*, vol. 12, p. 800, 1976.

[48] E. London and J. Coyle, *Eur. J. Pharmacol.*, vol. 56, p. 287, 1979.

[49] R. Squires and C. Braestrup, *Nature*, vol. 266, p. 732, 1977.

[50] C. Paden, B. McEwen, J. Fishman, L. Snyder and V. De Groff, *J. Neurochem.*, vol. 39, p. 512, 1982.

[51] W.L. McGuire, G.C. Chamness, M.E. Costlow and K.B. Horwitz, In: *Hormone-Receptor Interaction: Molecular Aspects*, G. Levey, Ed., New York: Marcel Dekker, 1976, p. 265.

[52] D. Houston and A. Howlett, *Mol. Pharmacol.*, vol. 43, p. 17, 1993.

[53] N. Zisapel and M. Laudon, *Brain Res.*, vol. 272, p. 378, 1983.

[54] L. Chatten and L. Harris, *Analyt. Chem.*, vol. 34, p. 1495, 1962.

[55] S. Gocan, G. Cimpan and J. Comer, In: *Advances in Chromatography*, E. Grushka and N. Grinberg, Eds., Boca Raton: Taylor & Francis, 2005, p. 113.

[56] R. Barlow, *Biochem. Soc. Sympos.*, vol. 19, p. 46, 1960.

[57] E. Vul'fius, O. Iurchenko and E. Zeimal', *Doklady Academii Nauk SSSR*, vol. 186, p. 1445, 1969 (in Russian).

[58] F. Dukhovich, M. Darkhovskii, E. Gorbatova and V. Kurochkin, Molecular Recognition: Pharmacological Aspects, New York: Nova Science Publishers, 2004.

[59] I. Wilson, *J. Biol. Chem.*, vol. 197, p. 215, 1952.

[60] I. Wilson and F. Bergmann, *J. Biol. Chem.*, vol. 185, p. 479, 1950.

[61] D. Pressman and M. Siegel, *J. Am. Chem. Soc.*, vol. 75, p. 686, 1953.

[62] D. Pressman, A. Grossberg, L. Pencl and L. Pauling, *J. Am. Chem. Soc.*, vol. 68, p. 250, 1946.

[63] A. Warshel and M. Levitt, *J. Mol. Biol.*, vol. 103, pp. 227–249, 1976.

[64] A. Warshel and R. Bora, *J. Chem. Phys.*, vol. 144, p. 180901, 2016.

8 Bound Water Effects on the Interaction of Chemical Groups

Formation of a stable ligand–receptor complex is possible only when complete displacement of water molecules occurs from the contact area. The role of the desolvation free energy barrier in ligand binding has been reliably confirmed through computer simulations [1, 2]. Nondisplaced (bound or conserved) water molecules increase not only the distance between the interacting ligand and its receptor but also the local dielectric constant of the medium, impeding complex formation resulting from decreased electrostatic interaction energy. Identification of stable and nonstable hydration shells in protein-binding sites can aid in designing selective drugs targeted to areas wherein water displacement occurs easily [3, 4, 5]. Another water-related issue is that the strength of hydration shells of different molecules and their molecular groups vary. For example, water molecules have a stronger binding to charged groups than to polar or nonpolar groups.

The enthalpic effect of water can be evaluated by comparing the potential energies of the direct binding of various groups in the drug molecule to the receptors with the hydration energies of these groups. Interactions between groups of the same nature and between groups of different nature are considered in the following manner. Interactions among charged groups is used as an example of interactions among the same nature groups, while interactions of charged groups with polar groups are used as an example of the interaction among groups of different nature.

8.1 POTENTIAL ENERGY OF INTERACTION BETWEEN A DRUG WITH CHARGED GROUPS AND ITS RECEPTOR WITH CHARGED GROUPS

The potential energy of interaction between the cationic groups $-NH_3^+$ and $-NH(CH_3)_2^+$ often found in drug molecules and carboxylate anion representing the anionic group in the active site of many classes of neuroreceptors [6] was calculated, and the results are presented in Table 8.1. Equilibrium distance (d_e) was calculated as the sum of van der Waals radii of the interacting groups (see Chapter 1). The d_e values correspond to the maximum energy of complex formation.

Although electrostatic interaction between ions is the most important type of interaction, several other interactions such as polarization and dispersion accompany it and are further considered in all cases, despite being of lesser importance.

8.2 POTENTIAL ENERGY OF INTERACTION OF CHARGED GROUPS WITH WATER

The results of calculations of the potential energy of hydration of the ions E_{i-w} with molar ratio "ion–water" 1:1 are presented in Table 8.2. Removal of two of these water molecules bound most strongly to the respective charged group (one per interacting ion) is sufficient for realization of the direct contact

DOI: 10.1201/9781003366669-9

between the charged groups at van der Waals distances. Hydration energy of the second layer is not considered because of its relatively small contribution [7]. ... It is assumed that the ion charge center and the charges of water dipole are on the same straight line. Dielectric permittivity is considered equal to 1.

The data presented in Tables 8.1 and 8.2 clearly suggest the displacement of the bound water molecules and unobstructed direct contact of the cationic group with the anionic group.

8.3 POTENTIAL ENERGY OF INTERACTION OF THE CHARGED GROUPS WITH POLAR GROUPS

Interaction of the groups of different nature is considered to be of, for example, distant ion–dipole interactions. It is assumed that the polar groups are connected to the carbon atom in the molecule; hence, refraction of the terminal bonds was estimated as ½ of the refraction of the C—C or N—C bond and added to the refraction of the main group. Dipole moments of the groups are presented in previous studies [8, 9]. It is assumed in calculations that the centers of charges of interacting groups were located on the straight line and that $\varepsilon = 1$. An example of detailed energy calculation is given for the interaction of the amide group of the peptide $-NH-\underset{\overset{\|}{O}}{C}-$ bond with the charged groups $-NH\left(\overset{+}{C}H_3\right)_2$, $-NH_3^+$, and $-COO^-$.

The charge at the amide group poles is 0.39 ē, and its dipole moment is $\mu = 3.7$ D [10]. Based on these values, the dipole length is $l = 1.98$ Å. Consequently, the equilibrium distance on the direct contact of this group with the $-NH(CH_3)_2$ group is $d_e = 5.56$ Å (from the side of the nitrogen atom ion, which is taken as the center of the ion charge). The equilibrium distance is evaluated from the following components: 3.17 Å (van der Waals radius of the charged group), 0.99 Å (dipole center), and 1.40 Å (van der Waals radius of the oxygen atom). The equilibrium distances between the considered amide group and the charged groups $-\overset{+}{N}H_3$ and $-COO^-$ are 4.09 and 4.54 Å, respectively. It is sometimes difficult to determine the complex structure because of which the following estimates of the d_e values from their van der Waals radii were used for other polar groups: 4 Å for interaction with $-\overset{+}{N}H_3$ and $-COO^-$ ions and 5 Å for interaction with the $-\overset{+}{N}H(CH_3)_2$ cation.

The values of potential energy E_{i-d} of the direct interaction of charged groups with the polar groups at a molar ratio of 1:1 are provided in Table 8.3.

TABLE 8.1
Potential Energy of Interaction of Cationic Groups with Carboxylate Anion (E_{i-i})

Ion Group	—COO⁻	
	d_e, Å	E_{i-i}, kcal/mol
-NH₃⁺	3.85	−82.2
-NH(CH₃)₂⁺	5.32	−58.4

TABLE 8.2
Potential Energy of Interaction of Charged Groups with Water E_{i-w} at Molar Ratio Ion–Water (1:1)

Charged group	d_e, Å	$-E_{i-w}$, kcal/mol
—COO⁻	3.82	8.36
—NH₃⁺	3.37	11.32
—NH(CH₃)₂⁺	4.84	5.06

8.4 POTENTIAL ENERGY OF INTERACTION OF POLAR GROUPS WITH WATER

The values of hydration potential energy of the polar groups E_{d-w} are presented in Table 8.4. The charge poles of the polar groups and those of water molecules are assumed to be positioned in-line (head-to-tail). Molar ratio "polar group–water molecule" is 1:1. The d_e value during contact of the amide group with the water molecule is 4.06 Å, wherein one half of the value (2.39 Å) is associated with the polar group and the other half (1.67 Å) is associated with the water molecule. For other dipole groups, an average value of 3.5 Å is used as the d_e value resulting from their van der Waals radii.

TABLE 8.3
Potential Energy of Binding of Charged Groups to Polar Groups, E_{i-d}

Dipole \ Ion	$-E_{i-d}$, kcal/mol		
	—COO⁻	—NH₃⁺	—NH(CH₃)₂⁺
—NH—C(=O)—	11.47	14.59	7.55
—CH₂OCH₂—	7.6	7.2	4.6
—CH₂OH	8.0	8.0	4.8
—CH₂C(=O)CH₂—	12.0	10.0	8.5
—C(=O)—O—	8.1	8.1	5.0
—CH=O	11.3	11.3	6.5
H₂C=O	10.8	11.8	6.7

TABLE 8.4
Potential Energy of Interaction of Polar Groups with Water, E_{d-w}. Ratio of Polar Group:H_2O = 1:1

Polar Group	$-E_{d-w}$, kcal/mol
—NH—C(=O)—	3.32
—CH₂OCH₂—	2.55
—CH₂OH	1.63
—CH₂C(=O)CH₂—	4.44
—C(=O)—O—	2.42
—CH=O	2.94
H₂C=O	3.07

8.5 EFFECT OF BOUND WATER ON BINDING

Table 8.5 shows a comparison of the total energy of hydration E_w of an ion–dipole interaction with the energy E_{i-d} of the direct binding of the ion–dipole interaction. The data provide information on whether or not the process of complex formation between the ligand molecule with polar groups (dipole) and the receptor charged residue (or vice versa, charged ligand with the receptor's polar residue) is favorable. The difference between the presented parameters defines the possibility of binding between charged and dipole groups.

The results of calculations presented in Table 8.5 reveal that the binding between charged groups and polar groups is unfavorable because the energy of hydration is higher than that of the direct interaction of the groups. This conclusion does not imply that the polar group cannot be completely associated with the charged group. Displacement of the bound water molecules can occur at the expense of interaction with the other groups, but such interaction causes significant penalty on the complex stability, which should be compensated somewhere else.

The interacting groups are surrounded by a higher number of water molecules than that in our model containing only two water molecules. Addition of more water molecules associated with charged and polar groups could only make their contact more unfavorable. The reason is that E_{i-d} is a constant for the pair of groups, while the E_w value, as a desolvation penalty, will increase with the addition of more water molecules (in absolute value). In essence, our model consistently characterizes the interaction of charged groups with polar groups in an aqueous medium.

8.6 CHANGES IN FREE ENERGY (ΔG) OF INTERACTION OF SAME-NATURE AND DIFFERENT-NATURE GROUPS

The conclusion on the important role of bound water molecules in molecular recognition is corroborated by experimental data presented in Table 8.6.

TABLE 8.5
Potential Energy E_{i-d} of Complex Formation between the Charged Group and the Polar Group and Their Individual Hydration Energy, E_w

Dipole \\ Ion	—COO⁻			—NH₃⁺			—NH(CH₃)₂⁺		
	$-E_{i-d}$	$-E_w$	$E_{i-d}-E_w$	$-E_{i-d}$	$-E_w$	$E_{i-d}-E_w$	$-E_{i-d}$	$-E_w$	$E_{i-d}-E_w$
—NH—C(=O)—	11.47	11.66	0.19	14.59	14.64	0.05	7.55	8.38	0.83
—CH₂OCH₂—	7.6	10.9	2.2	7.2	13.9	6.7	4.6	7.6	3.0
—CH₂OH	8.0	10.0	2.0	8.0	13.0	5.0	4.8	9.9	5.1
—CH₂C(=O)CH₂—	12.0	12.8	0.8	10.0	15.8	5.8	8.5	9.5	1.0
—C(=O)—O—	8.1	10.8	2.7	8.1	13.7	5.6	5.0	7.5	2.5
—CH = O	11.3	11.3	0.9	11.3	14.3	3.0	6.5	8.0	1.5
H₂C = O	10.8	11.4	0.6	11.8	14.4	2.6	6.7	8.1	1.4

Values are expressed in kcal/mol.

TABLE 8.6
Affinities of Indole Derivatives for Melatonin Receptors [11]

Compound	K_d, nM	Compound	K_d, nM
Melatonin	6.3	N-Acetyltryptamine	1600
6-Hydroxymelatonin	74	Tryptamine	>100 000
6-Methoxymelatonin	460	L-tryptophan	>100 000
2-Iodomelatonin	2.5	N-acetyltryptophan	>100 000
6,7-Dichloro-2-methylmelatonin	10	5-Methoxy-indole-3-acetic acid	>100 000
N-Acetylserotonin	3000	Serotonin	>100 000

The table shows the data on the binding of indole derivatives to the melatonin receptor, with the absence of charged groups in the melatonin orthosteric binding pocket [12]. The same-nature groups can be represented by the dipole acetamide group of the ligands bound to the receptor's dipole aceta-mide group of the Gly172 backbone. However, the receptor's dipole group might be another residue positioned closely to Gly172 in the binding pocket. The different-nature groups can be represented by the side chain or backbone of some uncharged (polar) residue of the receptor-binding pocket and charged groups of the ligands. Introduction of charged groups to the melatonin molecule and its analogs caused a sharp drop in the binding affinity, which becomes vanishingly small ($K_d > 100$ 000 nM). Charge neutralization by N-acetylation of the charged group makes the ligand adopted to the melatonin receptor; therefore, the affinity for N-acetylserotonin and N-acetyltryptamine is considerably higher than that for their charged counterparts serotonin and tryptamine, respectively. Affinity for N-acetyl-tryptophan remains extremely low because of the presence of the negatively charged carboxylic group. Introduction of uncharged substituents to the melatonin molecule affects the affinity only slightly.

Table 8.6 demonstrates that not only the binding of different-nature groups is ruled out but also these groups interfere with the binding of whole molecules mainly because of hindered desolvation.

Overall, the data on the binding of indole derivatives to the melatonin receptor presented in Table 8.6 are even more convincing than the data presented in Table 8.5. K_d values are associated with changes in free energy ΔG comprising all factors of the compounds binding with the receptors. However, the computational model is useful for the semi-quantitative assessment of the effect of different ligand groups on complex formation.

Thus, the different nature of binding of the same-nature and different-nature groups of drug molecules and receptors should be adequately estimated.

For investigating the binding of different-nature groups, only the interaction of ion and dipole groups is considered. However, it can be argued that the interaction of other different-nature groups "ion–nonpolar group" and "dipole–nonpolar group" would also be unfavorable.

For example, in a previous study [13], a simple physical model was constructed for estimating the interaction energy of a ligand and a protein binding center in an aqueous medium, depending on whether the protein group and the ligand have charged groups or nonpolar groups. The calculated values of binding free energy show the advantage of the combinations "nonpolar ligand–nonpolar protein group" and "charged ligand–oppositely charged protein group." Almost complete disadvan-tage of the interaction of nonpolar groups with charged groups occurs due to hydration effects.

By contrast, we cannot exclude the fact that the ionic group of a ligand has contact with nonpolar groups in case of prior water displacement from the contact pocket resulting from ion–ion contact. For example, it is well known that the cationic group of acetylcholine and nicotine comes into con-tact with tryptophan aromatic groups in the active site of the nicotinic acetylcholine receptor, where the cation-π complex is formed [14, 15].

Bound water is one of the main factors for the recognition ability of a drug toward specific receptors. By lowering the binding enthalpy and reducing the lifetime of drug complexes with non-specific acceptors, the bound water thus prevents nonspecific binding of drugs in an organism, con-tributing to the recognition of specific receptors.

REFERENCES

[1] R. Dror, A. Pan, D. Arlow, D. Borhani, P. Maragakis, Y. Shan, H. Xu and D. Shaw, *Proc. Natl. Acad. Sci. USA*, vol. 108, p. 13118, 2011.

[2] J. Mondal, R. Friesner and B. Berne, *J. Chem. Theory Comput.*, vol. 10, p. 5696, 2014.

[3] S. Vukovic, P. Brennan and D. Higgins, *J. Phys.: Condens. Matter*, vol. 28, p. 344007, 2016.

[4] F. Spyrakis, M. Ahmed, A. Bayden, P. Cozzini, A. Mozzarelli and G. Kellogg, *J. Med. Chem.*, vol. 60, p. 6781, 2017.

[5] D. Robinson, W. Sherman and R. Farid, *ChemMedChem*, vol. 5, p. 618, 2010.

[6] A. Venkatakrishnan, X. Deupi, G. Lebon, C. Tate, G. Schertler and M. Babu, *Nature*, vol. 494, p. 185, 2013.

[7] J. Webb, Enzyme and Metabolic Inhibitors, vol. 1, New York: Academic Press, 1966.

[8] E. Moelwyn-Hughes, *Proc. Cambridge Phil. Soc.*, vol. 45, p. 477, 1949.

[9] V. Minkin, O. Osipov and Y. Zhdanov, Dipole Moments in Organic Chemistry, Leningrad: Khimiya, 1968, pp. 73–77 (in Russian).

[10] C. Cantor and P. Schimmel, Biophysical Chemistry, vol. 1, New York: W.H. Freeman and Co., 1980, pp. 245–247.

[11] M. Dubocovich and J. Takahashi, *Proc. Natl. Acad. Sci. USA*, vol. 84, p. 3916, 1987.

[12] B. Stauch, L. Johansson, J. McCorvy and N. Patel, *Nature*, vol. 569, p. 284, 2019.

[13] R. Baron, P. Setny and J. McCammon, *J. Am. Chem. Soc.*, vol. 132, p. 12091, 2010.

[14] S. Sine, P. Quiram, F. Papanikolau, H. Kreienkamp and P. Taylor, *J. Biol. Chem.*, vol. 269, p. 8808, 1994.

[15] W. Zhong, J. Gallivan, Y. Zhang, H. Lester and D. Dougherty, *Proc. Natl. Acad. Sci. USA*, 95, p. 12088, 1998.

9 Information Required for Receptor Recognition

9.1 PHARMACOPHORE AND ITS ELEMENTS

Drugs selective to a certain receptor must have common structural features. A set of such features allowing recognition of the receptor is termed as a pharmacophore. Defining a pharmacophore is often challenging, as, in many cases, the compounds recognizing predominantly the same receptor have essentially different chemical structures. The more the pharmacophore elements are in the compound composition, the higher is the possibility of being selective. Among all pharmacophores existing in a molecule, only one conformation is associated with maximum activity—bioactive conformation, which is the one conformation the molecule assumes during interaction with the receptor. This conformation is not necessarily the most stable one for the free molecule. A rigid molecule displays a significantly higher activity than its conformationally flexible analog because of the lower energy consumption needed for conformational changes on binding to the receptor.

To date, the protein structure as well as localization and identification of the active (binding) site has been determined for several neuronal receptors (nicotinic acetylcholine receptor, α_1-adrenergic receptor, muscarinic acetylcholine receptor, to name a few) [1, 2]. However, these data may be not sufficient for determining the possibility of the compound binding because, as a rule, conformational mobility of the residues in the receptor active site is unknown. The pharmacophore contains information on the receptor active site based, in most cases, on the principle of ligands' complementarity. Complementarity is defined by steric and electrostatic factors. The first factor considers active site topography—a molecule must have the corresponding geometry to be bound. The steric factor also includes stereoisomerism of the molecule. The second factor is charge distribution in the molecule presuming mirrored charge distribution in the receptor.

Information on the receptor binding site can also be obtained indirectly using a three-dimensional pharmacophore containing information on the spatial location of the interacting groups in the ligand. A two-dimensional (2D) pharmacophore defining only the distances between the groups without characterization of their spatial organization is much less informative.

In case when the receptor has several binding sites, there could be more than one variant with complementary arrangement; hence, there could be more than one corresponding pharmacophore. The drug molecule can occupy the mediator site either completely or partially or can hinder this from the distance according to the principle of allosteric interaction (Figure 9.1).

However, the same molecule can display affinity to several receptors assuming it has various bioactive conformations. For example, the conformationally flexible neuromediator acetylcholine can adjust itself to either the muscarine- or nicotine-acetylcholine receptor. Moreover, active conformations of the mediator for these two receptors are different.

DOI: 10.1201/9781003366669-10

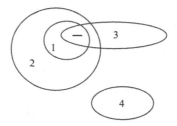

FIGURE 9.1 Hypothetic schema of the receptor sites expressing affinity to various ligands.

1 – mediator site, 2, 3 – ligand sites of competitive type, 4 – ligand sites of noncompetitive type. The anionic receptor site is marked with a minus symbol.

To simplify pharmacophore identification significantly, conformational analysis can be limited to a dozen or hundreds of conformers. This is precisely the case considered in classic pharmacophore studies conducted by Kier [3, 4, 5, 6, 7]. He demonstrated pharmacophore identification using a set of relatively small mAChR agonist molecules, wherein heteroatoms in acetylcholine **10.1**, muscarine **9.1**, and muscarone **9.2** are aligned in the same positions in space. However, it is difficult to align another selective mAChR agonist, namely, oxotremorine **9.5**, with this set of compounds. Kier showed [5] that the protonated form of oxotremorine in one of the thermodynamically preferable conformations corresponds to the pharmacophore shown in Figure 9.2A, where the triple bond displaying excessive electron density similarly to that of the oxygen atom occupied the position of the ester oxygen atom.

Pharmacophores of histamine [8], 5-hydroxytryptamine (5-HT; serotonin) [9, 10, 11, 12], adrenergic [13, 14], and other receptor ligands were elucidated based on the analysis of antagonist molecule structures (see also [15]). The pharmacophore of the serotonin receptor suggested in [6] consists of atoms carrying the most distinct charges (Figure 9.2 D).

The pharmacophore of the histamine receptor ligand (Figure 9.2 C) contains two possible arrangements of atoms corresponding to two different histamine conformations.

As evident from Figure 9.2, the pharmacophores are generated using different approaches. Some pharmacophores are designed based on the locations of key charges, whereas the others consider nonpolar radicals. Several pharmacophores are suggested for the same ligands.

In some cases, a pharmacophore containing only two elements is presented as a scheme. Such a pharmacophore structure does not explain ligand selectivity, which is expressed in varying activities of stereoisomers. However, in the absence of chiral centers, the pharmacophore consisting of two elements can provide high selectivity. Such pharmacophores are suggested, for example, for GABA and hexamethonium (see Table 7.4).

The search for pharmacophores common for several receptor classes is underway. For example, the common pharmacophore of the ligands of 5-HT_2- and D_2-receptors shown in Figure 9.2 F has proven useful for explaining the low selectivity of neuroleptics.

The triangle-shaped pharmacophore is theoretically sufficient for the proper recognition of a receptor. The simple shape of the pharmacophore is preferable to avoid overlap of pharmacophores of the ligands of different receptors (see Figure 9.3).

9.2 PHARMACOPHORE GOVERNING RECOGNITION OF mAcH RECEPTORS

Using highly selective mAChR ligands as an example, we consider what features of the molecules control their selectivity. The Kier scheme (Figure 9.2 A) conveys the similarity in the positions of key atoms in the acetylcholine, muscarine, and muscarone molecules. It is generally accepted that oxygen atoms in the ligands form hydrogen bonds with the receptor, which ensures the stability of

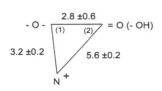

A Pharmacophore of muscarinic receptor ligand [3];

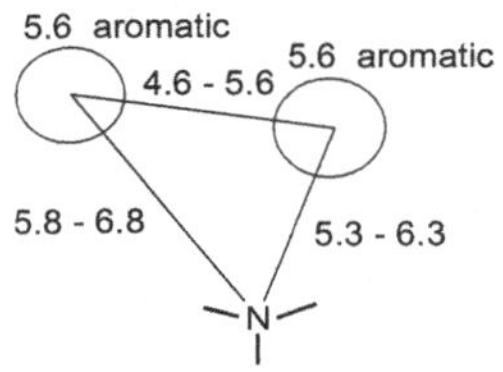

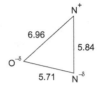

B Pharmacophore of H₁-receptor antagonist [8];

C Pharmacophore of histamine receptor ligand [5];

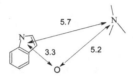

D Pharmacophore of serotonin receptor ligand [6];

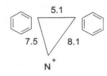

E Pharmacophore of 5-HT₃-receptor antagonist [11]

F   Pharmacophore common for 5-HT₂ and D₂-receptor ligands [16].

FIGURE 9.2   Pharmacophores of the ligands of various receptors. Distances are expressed in Å.

the complex [17, 18, 19]. However, a comparison of the structures of numerous mAChR receptor ligands raises one question: whether the oxygen atoms, which play the main role, at positions **1** and **2** in the Kier scheme participate in the formation of hydrogen bonds. As exemplified by the compounds depicted in Figure 9.4, heteroatoms or even multiple bonds, as well as either proton-donor or proton-acceptor atoms, can be present at these positions[1].

Analysis of the charge distribution in the molecules of mAChR ligands demonstrated that the three most pronounced charges occupy similar positions. Vertices 1 and 2 of the triangles in the Kier's scheme represent (δ^-) poles of the respective dipoles [3]. In some compounds, the negative charge is not localized on the oxygen atom. The elevated electron density is concentrated on the carbon atoms of oxotremorine, iperoxo, hexbutenol, and hexbutinol in sp²- and sp-hybridization. In these cases, multiple bonds functionally replace the ester oxygen atom.

Table 9.1 lists the calculated distances between the cationic group's nitrogen atom and either oxygen atoms (1) and (2) or multiple bonds for the compounds shown in Figure 9.4 in their most stable conformation. The nitrogen atom (in the ammonium or protonated amine form) is treated as a charge center of the cationic group. Summarizing the data, the ligands of muscarinic receptors

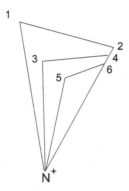

FIGURE 9.3   Comparison of pharmacophores of the ligands of different receptors:

N^+,1,2 - pharmacophore of the serotonin receptor ligand

N^+,3,4 - pharmacophore of the 5-HT$_3$-receptor antagonists

N^+,5,6 - pharmacophore of the mAChR ligand.

(the size of these pharmacophores is shown in Figure 9.2-A, D, E).

are rather conserved with regard to localization of the most pronounced charges, and variable with respect to the atoms carrying these charges.

The three atoms (or groups) discussed above that possess the most distinct charges display the most long-range activity in the composition of the muscarinic receptors. They play a main role in maintaining the correct orientation of the molecule on the receptor. As shown in Chapter 7, electrostatic interactions are essential at this stage. The association of the compound with mAChR does not depend on affinity, but rather, it is defined by the availability of key charges located at certain distances from each other. The average distances listed in Table 9.1 for the compounds with various structures are close to the distances reported by Kier (see Figure 9.2 A).

9.3   RECOGNITION SIGNATURE

In essence, recognition is the phenomenon based on the information encrypted in the structure of the compound and its receptor. As reported by Quastler [28], in most cases, not the entire information contained in the molecular object is used, but rather, its certain part is termed as "signature." The signature is responsible for certain function of the molecule not allowing "distraction" (reaction) to factors interfering with this function. On completion of the function, another signature that is responsible for another property of this molecule could be put first. Functionally, any drug molecule includes at least two signatures: with one signature consisting of a small number of atoms carrying pronounced charges that are responsible for recognition, and another signature that is responsible together with the first one for complex stability. Let us term the signature responsible for selectivity as "recognition signature" when compared with the pharmacophore that could be constructed based on the principle of the highest activity. Atoms marked as N1 and N2, which are listed in Table 9.1, form the recognition signature of mAChR ligands. It is likely that the pharmacophores described in literature not always coincide with the recognition signatures based on the long-range action of its elements.

High recognition ability is necessary but not sufficient for demonstrating high activity with respect to this receptor. For example, various nonpolar groups not present in the composition of the recognition signature impart activity to the muscarinic receptor ligands. The nonpolar and slightly

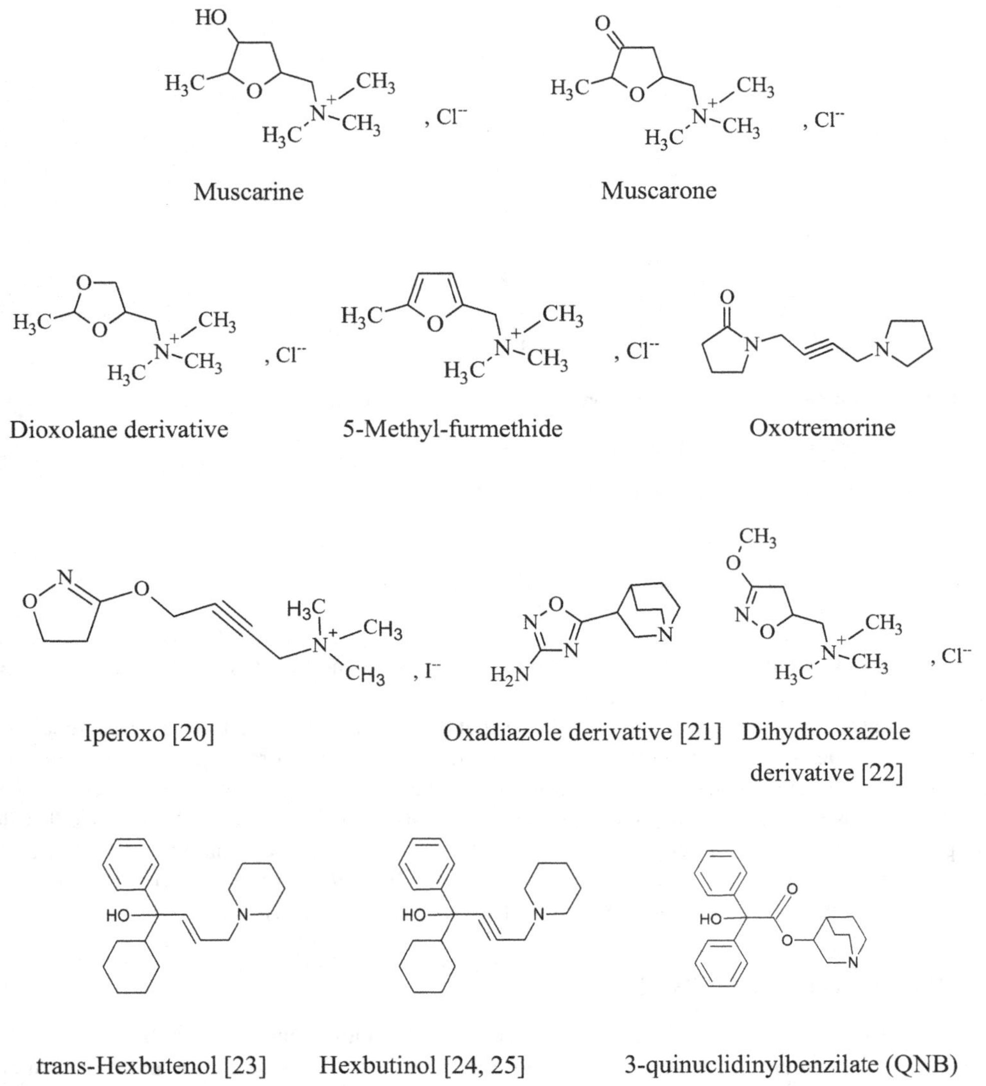

FIGURE 9.4 Structure of mAChR ligands.[1]

polar groups cannot significantly enhance the recognition ability of the compounds but could reduce its ability to significantly create steric hindrance for the "correct" orientation of the molecule on the receptor. Complementarity of the compound and receptor molecules without doubt is the basis for molecular recognition. However, this principle can be clarified in the following way. The recognition ability of the compound is controlled by the complementarity of the recognition signature, but activity (affinity) is controlled by the complementarity of the entire molecule. As elaborated below, the increase in the affinity of the compound for the receptor could be the reason for the reduced selectivity in the case, when it is achieved through the contact groups not present in the composition of the recognition signature. The charge-based triangle-shaped recognition signature facilitates unique fixation of the molecule on the specific receptor. The compounds with a recognition signature involving many elements could be less selective because of the overlapping of elements from

TABLE 9.1

Distances (in Å) between the Key Charges * in Compounds Shown in Figure 9.4

Compound	Distances and Charge Positions		
	N$^+$ - (1)**	N$^+$ - (2)**	(1) – (2)
Muscarine	N$^+$ - (-O-)	N$^+$ - O (OH)	O – O
	2.9	5.4	3.5
Muscarone	N$^+$ - (-O-)	N$^+$ - (= O)	O – O
	3.1	5.5	3.5
Dioxolane derivative	N$^+$ - O (1)	N$^+$ - O (2)	O – O
	2.9	3.9	2.2
5-Methyl-furmethide	N$^+$ - O	N$^+$ - =	O - =
	3.3	4.6	2.2
Oxotremorine	N$^+$ - ≡	N$^+$ - O	≡ - O
	2.4	5.3	3.2
Iperoxo	N$^+$ - ≡	N$^+$ - O (C-O-C)	≡ - O
	3.2	6.1	3.0
Oxadiazole derivative	N$^+$ - O	N$^+$ - N(=)	O - N(=)
	3.9	4.9	2.3
Dihydrooxazole derivative	N$^+$ - (- O -)	N$^+$ - O (OCH$_3$)	O - O (OCH$_3$)
	2.9	5.5	3.5
trans-Hexbutenol	N$^+$ - =	N$^+$ - O	= - O
	2.6	5.2	2.8
Hexbutinol	N$^+$ - ≡	N$^+$ - O	≡ - O
	2.5	5.6	3.4
QNB	N$^+$ - (- O -)	N$^+$ - (= O)	O – O
	3.5	5.2	2.3
Average	3.0 ± 0.4	5.6 ± 0.4	2.9 ± 0.5
Average, by Kier	3.2 ± 0.2	5.6 ± 0.2	2.8 ± 0.6

  * Charge distribution was calculated using the AM1-BCC method [26] with Chimera software [27]
 ** Positions (1) and (2) are shown in Figure 9.2A

different signatures. The compounds that do not have a complete set of signature elements can also display lower selectivity. For example, promazine (Figure 7.6) and cyproheptadine (Figure 5.3) are designed mainly based on the "anchor group" principle. This is the reason why they demonstrate low selectivity (see Table 7.4). Even slight changes in the recognition signature affect the activity profile, as indicated by the data shown in Table 9.2. The selective antagonists of muscarinic receptors— amino esters of the substituted glycolic and acetic acids—were used as test compounds, and nicotinic receptors and butyrylcholinesterase (BChE) were used in addition to mAChR as complex-forming acceptors. The hydroxyl group in amino glycolates C-**4.11**, C-**4.43**, **QNB**, and benactyzine forms an intramolecular hydrogen bond with the oxygen atom in the carbonyl group, thus limiting conformational mobility of the molecule. Transition to the respective acetic acid aminoesters **C-4.42, C-5.1, C-6.10,** adiphenine, and aprophene without the hydroxyl group removes this restriction, and in all cases, it increases activity toward nonspecific acceptors for all compound pairs "glycolate–acetate" almost identically: fourfold for the nicotinic receptors and fivefold for butyrylcholinesterase. Thus, targeted control of the affinity ratio to various receptors is possible for a certain series of compounds according to standard structure modification.

Considering that new mediators and their corresponding new receptors are being discovered continuously, it is evident that there are unknown receptors and unknown binding sites. By determining recognition signatures specific for known receptors, designing virtual signatures (by searching

TABLE 9.2
Affinity of Amino Esters of the Substituted Glycolic and Acetic Acids to mAChR, nAChR, and Butyrylcholinesterase (BuChE) [29]

Name	Compound Structure	K_d, M		
		mAChR	nAChR	BuChE
C-4.11		$5.6 \cdot 10^{-11}$	$2.8 \cdot 10^{-5}$	$7.4 \cdot 10^{-6}$
C-4.42		$6.1 \cdot 10^{-10}$	$9.7 \cdot 10^{-6}$	—
QNB		$1.4 \cdot 10^{-10}$	$1.2 \cdot 10^{-5}$	$7.7 \cdot 10^{-6}$
C-5.1		$2.0 \cdot 10^{-10}$	$3.2 \cdot 10^{-6}$	$1.4 \cdot 10^{-6}$
C-6.16		$1.1 \cdot 10^{-10}$	$1.6 \cdot 10^{-5}$	$5.4 \cdot 10^{-6}$
C-6.10		$9.2 \cdot 10^{-11}$	$4.1 \cdot 10^{-6}$	$1.1 \cdot 10^{-6}$
Benactyzine		$2.9 \cdot 10^{-9}$	$1.9 \cdot 10^{-5}$	$6.2 \cdot 10^{-6}$
Adiphenine		$3.4 \cdot 10^{-8}$	$3.5 \cdot 10^{-6}$	$1.1 \cdot 10^{-6}$
Aprophene		$7.6 \cdot 10^{-9}$	—	$2.8 \cdot 10^{-6}$

omissions) can be attempted. The number of alternatives would likely not be very large, as the receptor's active sites have a relatively small size of approximately 100 Å^2 (relevant estimations can be found at the beginning of Chapter 7).

The signatures of agonists are, as a rule, either identical or very close to the signature of endogenous effectors. However, such similarity is not required for antagonists. The structures of neuroleptics belonging to the phenothiazine family, blockers of D_2-dopamine receptor (chlorpromazine and fluphenazine), and butyrophenone derivatives (haloperidol and droperidol) are different from dopamine and between themselves.

If the goal is to improve a certain property (affinity or selectivity) of the compound, then the most comprehensive approach to the design of the compound should include the identification of the structural elements of the molecule that must not be changed and the ones that could be changed, and what new substituents could be introduced. To preserve specificity toward the target receptor, the recognition signatures should not be changed. Moreover, the positions of key charges should not be changed, while the specific atoms or groups carrying these charges may vary. It might be acceptable to vary the nonpolar and slightly polar groups. On changing the compound's structure to enhance the activity of the compound, partial binding free energy of new substituents can be estimated *in silico* (for details, see Chapter 4), given that the receptor protein structure is known.

The most typical feature of the compounds that have one ionic group in the recognition signature is that the molecule must not have any other ionic group. Otherwise, the recognition algorithm would be disrupted, which could result in reduced selectivity. However, acetylcholine and bis-symmetrical nicotinic acetylcholinergic molecules contain one and two cationic groups, respectively. Thus, it can be concluded based on the structures of suxamethonium, atracurium, d-tubocurarine, and other di-cationic species that the anionic groups of the nAChR active sites are separated by the relatively large distances and do not interfere significantly with the binding of small molecules of the mono-cationic type such as acetylcholine, tetramethylammonium, and others to the receptor. According to accumulating experimental evidence, at least two protein subunits of the nAChR molecule have binding pockets for ACh [30].

The molecule of a therapeutic compound can be considered as a minicomputer performing a recognition program that is encrypted in the molecule structure, starting from the long-range stage and

TABLE 9.3
Profile of the Affinity of 5-HT$_2$-Receptor Ligands. Relative K_d Values Are Shown in Parentheses

Compound	K_d, nM					
	5-HT$_2$	D$_1$	D$_2$	α_1	α_2	H$_1$
Metergoline	0.8 [32]	22 [33]	23 [34]	38 [34]	380 [34]	1100 [34]
	(1)	(28)	(29)	(48)	(475)	(1375)
Metitepine	1.9 [32]	210 [35]	4.0 [34]	0.47 [34]	48 [34]	4.9 [34]
	(1)	(111)	(2.1)	(0.25)	(25)	(2.6)
Cyproheptadine	6.5 [32]	114 [36]	31 [34]	100 [34]	760 [34]	2.7 [34]
	(1)	(18)	(4.8)	(15)	(117)	(0.42)
LSD	8.2 [32]	30 [35]	20 [35]	160 [34]	58 [34]	1300 [37]
	(1)	(3.7)	(2.4)	(20)	(7.1)	(159)
Methysergide	12 [32]	290 [35]	200 [34]	2300 [34]	2600 [34]	33 000 [38]
	(1)	(24)	(17)	(192)	(217)	(2750)
Cinanserine	41 [32]	2790 [36]	1600 [34]	1200 [34]	>1000 [34]	1200 [34]
	(1)	(68)	(39)	(29)	(>24)	(29)
Serotonin	296 [39]	44 000 [40]	43 000 [41]	23 000 [42]	>100 000 [43]	>46 000 [38]
	(1)	(149)	(145)	(78)	(>338)	(>155)

ending with the contact stages with specific and nonspecific acceptors. Furthermore, the algorithm of the recognition program includes mechanisms protecting the drug from nonspecific binding.

9.4   SELECTIVITY AND AFFINITY: ARE THESE PARAMETERS ALWAYS RELATED?

The selectivity of drugs implies the ability to bind to one receptor at a higher degree than to the other ones. The definition for selectivity is described here in the context of the fact that the term "selectivity" is unjustifiably used to characterize the preferable binding of a compound to its receptor when compared with other compounds. In the latter case, the issue is the activity of the compound rather than its selectivity. The following concept was proposed: the degree of recognition specificity is symbatic to the degree of ligand binding; hence, it is expressed in terms of the binding free energy [31]. However, this is not generally true. The increase in affinity (decrease of K_d value) toward specific receptors indicates increased complementarity and, at first glance, should lead to increased selectivity. However, this relation is not always observed. The data presented in Tables 9.3 and 9.4 can serve as an example of absence of this relation.

The K_d values reported by different authors depend, to a certain extent, on experimental conditions: animal species, tested organ, labeled competing agent, temperature, and other factors. For comparison, we have used the data produced under identical or similar conditions.

The compounds presented in Table 9.3 are arranged in a descending order based on the affinity to the 5-HT$_2$-receptor. The most active ligand—metergoline—demonstrates higher selectivity than the less active metitepine, cyproheptadine, and LSD, but it is less selective than methysergide and serotonin. Serotonin recognizes only the 5-HT$_2$-receptors.

As shown in Table 9.4, fluphenazine, which is the most active compound toward D$_2$-receptors, demonstrates relatively low selectivity. In terms of selectivity, haloperidol and pimozide are significantly better than compounds that demonstrate either higher or lower affinity to the D$_2$-receptor. At the same time, chlorpromazine and clozapine, which are the most active compounds to the key receptor, display low selectivity. Modification of the pirenzepine molecule with the objective to improve its selectivity toward (M$_1$/M$_2$)-muscarinic acetylcholine receptor subtypes was studied in [61, 62, 63, 64, 65]. It was mentioned that the increase in affinity to the M$_1$-subtype was accompanied by the decrease in selectivity.

Affinity belongs to the category of objective parameters of the compound in relation to a certain receptor. Emergence of new information on affinity to the receptors not considered before can

TABLE 9.4
Affinity Profiles to the Receptors of Some Neuroleptics – Antagonists of D$_2$-Dopamine Receptors. Relative K_d Values Are Shown in Parentheses

Neuroleptic	K_d, nM					
	D$_2$	D$_1$	α_2	H$_1$	5-HT$_2$	mACh
Fluphenazine	0.5 [44]	5.0 [36]	580 [45]	15 [46]	2.5 [47]	680 [48]
	(1)	(10)	(1160)	(30)	(5.0)	(1360)
Pimozide	0.9 [49]	1400 [50]	1100 [51]	1100 [52]	8.0 [47]	1100 [53]
	(1)	(1556)	(1222)	(1222)	(8.9)	(1222)
Haloperidol	1.0 [54]	50 [36]	1800 [51]	2600 [52]	95 [55]	2700 [48]
	(1)	(50)	(1800)	(2600)	(95)	(2700)
Chlorpromazine	1.1 [56]	22 [36]	840 [51]	28 [52]	1.4 [57]	70 [58]
	(1)	(20)	(766)	(25)	(1,3)	(64)
Clozapine	6.9 [44]	170 [36]	4100 [59]	20 [52]	5.0 [47]	26 [60]
	(1)	(25)	(594)	(2,9)	(0,72)	(3,8)

change the notion of selectivity of this compound. Hence, the increase in affinity of the drug to the specific receptor not always results in the increase in selectivity.

9.5 RECOGNITION AND CONFORMATIONAL CHANGES IN DRUG AND RECEPTOR

Conformational changes in the drug molecule and its receptor during their interaction could be an important factor of molecular recognition. Active centers of receptors, enzymes, and other proteins are not always vacant—they are occupied by either buried water molecules, molecules of non-specific compounds, or both.

Koshland [64] introduced the "induced fit" hypothesis, according to which the substrate induces conformational changes in the respective enzyme, facilitating substrate binding and its orientation that promote the enzymatic reaction. The nonspecific (noncomplementary) substrate does not have the ability to induce such conformational changes in the active center. The Koshland's hypothesis, which was corroborated experimentally, was later applied to other protein–ligand associations including ligand–receptor interactions. It is likely that, in most cases, there is mutual adjustment between the ligand and the binding pocket in the receptor protein.

A review of the studies investigating conformational changes in enzymes under the action of low-molecular-weight ligands was conducted by Blumenfeld [65]. The study concluded that the induction of conformational changes is a vital step in the enzymatic process. The same conclusion is applicable to the ligand–receptor interactions.

Recently, a model of conformational selection was proposed not as an alternative but as a complement to the induced fit model [66]. In the unbound state, a native protein can exist in multiple conformational states with different degrees of affinity to a ligand. Conformational selection occurs when a ligand is bound to a certain protein conformation, thus producing the effect or blocking it. A kinetic scheme for the model with two conformational states of a receptor can be outlined as follows:

$$R^{s1} \rightleftarrows R^{s2}$$

$$L + R^{s1} \rightleftarrows LR^{s1}$$

where the R^{s1} is a strongly binding conformation and R^{s2} is a weakly binding conformation. The authors of [66] suggest that conformational selection model and the induced fit model might be just two limiting cases of the same process where protein conformational ensemble distribution is affected by ligand binding.

A kinetic binding model for conformational selection differs from the induced fit, and there are proposals on how to attribute the valid model for binding process description from chemical relaxation time [67]. Hammes et al. have proposed the flux of a reaction (the reactant concentration × the forward rate constant) for discovering the possible mechanism of conformational changes [68]. The importance of conformational transitions is reflected in both models, and extensive use of NMR spectroscopic methods to study the coupling of conformational states of proteins with ligand binding is proposed.

In one of the first studies [69, 70] on conformational changes of the nicotinic receptors from stingray electric organ,[2] the authors determined the fluorescence quantum yield of quinacrine bound to the receptors under the action of ligands. The agonists cause fast structural transitions in the receptor protein, while the antagonists affected quinacrine fluorescence to a lesser degree. The experimental results are described using a three-step model, with the first two steps being the fastest. Similar kinetics of the ligand binding to the stingray electric organ acetylcholine receptors is demonstrated by recording the intrinsic fluorescence of the receptor protein [71, 72].

The transitions of the T-cell receptor induced by the specific ligand are examined in [73]. Two stages of interaction are identified—the fast one (diffusion controlled) and the slow one, with the rate constant ranging from 2 to 4 s^{-1}. The authors suggested that the second step corresponds to the induced fit process (slow rearrangement of the protein structure caused by ligand binding).

The kinetics of conformational changes in the α_2-adrenoreceptors tagged with a fluorescent probe during binding of agonists and antagonists is investigated using the method of Förster resonance energy transfer in [74]. The study revealed the protein regions subjected to conformational changes under the action of agonists, antagonists, and partial agonists. The fluorescence intensity for the partial agonist clonidine is 1.5-fold lower than that for the agonist noradrenaline, which implies larger conformational changes in the protein induced by the pure agonist noradrenaline.

The binding kinetics of the anesthetics halothane, sevoflurane, and others with the ion channel formed by two $A\alpha_2$-L1M/L38M subunits is investigated in [75] also revealed two stages in the process of protein conformational changes—the fast stage with a rate constant of approximately 20 s^{-1}, and the slow stage with a rate constant of 1 s^{-1}. The slow stage corresponds to the conformational transition of the complex obviously induced by the ligand.

From our point of view, conformational changes in the ligand–receptor complexes can be classified into three types:

1) Fast conformational changes occurred during the orientation of a molecule on the receptor as the first step of complex formation (induced fit model according to Koshland). Such varying degrees of changes occur during the binding of many (if not all) ligands of that receptor.
2) Changes in the position of the antagonist on the receptor because of the protein's relatively slow motions (auto-oscillation processes). In the process, the K_d value can either decrease or increase, and in some cases, the equilibrium might not be found at all.
3) Changes in the characteristics of the hydrogen bond formed in the primary complex, or emergence of new hydrogen bonds; stability of the complex is increased in the process.

Three types of conformational changes can be considered as sequential steps of the process of a ligand binding to the receptor. The ligand structure (especially nonpolar groups) controls what stage the process of complex formation stops.

The rate of conformational changes of a ligand may be lower than the rate of the complex formation with the receptor. For this reason, a sterically acceptable complex may not be formed. Reference data on the rates of conformational changes of small-sized molecules are given in Table 9.5. The lifetime of individual conformational states is calculated based on the energy barriers of conformational transitions.

In proteins, conformational changes can be either small, such as the torsional rotation of a residue side chain, or large, such as α-helix space shift or unfolding. Presumably, local conformational changes in the protein have values of $\Delta G^{\#}$ comparable to or even smaller than those listed in Table 9.5. It is difficult to experimentally measure such conformational barriers, but there are some computational estimates. Specifically, there are both experimental and computational data for protein folding and unfolding transitions [76–80]. Free energy barriers for large-scale conformational transition from α-helix to β-sheet calculated from molecular simulations range from 5 to 10–20 kcal/mol [81, 82].

The conformational changes between a free receptor and a membrane-bound receptor may explain why K_d values measured in radioligand experiments using tissue homogenates and on "pure" cloned receptors may greatly differ from each other [92].

In conclusion, conformational changes in both the ligand and the receptor should always be treated as a recognition factor. It can significantly affect the binding of drugs to specific and nonspecific receptors.

TABLE 9.5
Characteristic Rate Constants of Conformational Transitions in Typical Organic Molecules

Transition	Transition Barrier $\Delta G^{\neq}$, kcal/mol[1]	Rate Constant k, s[-1][2]	Lifetime $\tau=1/k$, s
Internal rotation around single bonds			
CH_3CH_3	2.88 [83]	$4.9 \cdot 10^9$	$2 \cdot 10^{-10}$
CH_3NH_2	1.94 [83]	$2.4 \cdot 10^{10}$	$4 \cdot 10^{-11}$
CH_3NHCH_3	3.28 [83]	$2.6 \cdot 10^9$	$3.8 \cdot 10^{-10}$
$CH_3N(CH_3)_2$	4.41 [83]	$4 \cdot 10^8$	$2.5 \cdot 10^{-9}$
H_3C–N(CH$_3$)–C(=O)–t-Bu (amide)	12.2 [84]	$2.6 \cdot 10^4$	$3.9 \cdot 10^{-5}$
2,2'-dibromobiphenyl (Br Br)	19.0 [85]	$2.4 \cdot 10^{-2}$	40.9
o-bis[CH$_2$(t-Bu)]benzene	11.5 [86]	~33.6 [3]	$3 \cdot 10^{-2}$
Cycle inversion of the type X–Z ⇌ X–Z (cyclohexane, X=H)	~10 [87]	$4.3 \cdot 10^4$	$2.3 \cdot 10^{-5}$
HN–ring (H)	14.5 [88]	$1.6 \cdot 10^6$	$6.1 \cdot 10^{-7}$
CH_3N–ring (H)	14.4 [89]	$1.5 \cdot 10^4$	$6.5 \cdot 10^{-5}$
O–ring (H)	9.9 [90]	234.4	$4.3 \cdot 10^{-3}$
cyclophane inversion	>28 [91]	$<2.3 \cdot 10^{-11}$ [3]	$>4.3 \cdot 10^{10}$

[1] In some cases, the calculated barrier of the potential energy E is given.

[2] The constant k was calculated for experimental barriers using the equation $k = \chi \dfrac{k_B T}{h} e^{-\Delta G^{\neq}/RT}$, where k_B is the Boltzmann constant, and χ is the transmission coefficient, i.e., the fraction of the molecules changed into a new conformer after the barrier is reached, commonly assumed to be 1. For the calculated barriers (E), the rate constant was calculated using the equation $k = Ae^{-E/RT}$ at T=298 K.

[3] Assuming $A=10^{10}$ s^{-1}.

NOTES

1 References are provided for compounds that are not often cited in literature.
2 The tissues of stingray electric organ contain acetylcholine receptors at a high concentration, which helps their purification.

REFERENCES

[1] J. Shonberg, R. Kling, P. Gmeiner and S. Lober, *Bioorganic Med. Chem.*, vol. 23, p. 3880, 2014.

[2] T. Grutter and J.-P. Changeux, *Trends Biochem. Sci.*, vol. 26, p. 459, 2001.

[3] L. Kier, *Mol. Pharmacol.*, vol. 3, p. 478, 1967.

[4] L. Kier, *Mol. Pharmacol.*, vol. 4, p. 70, 1968.

[5] L. Kier, *J. Med. Chem.*, vol. 11, p. 441, 1968.

[6] L. Kier, *J. Pharm. Sci.*, vol. 57, p. 1188, 1968.

[7] L. Kier, *J. Pharm. Pharmacol.*, vol. 21, p. 93, 1969.

[8] P. Borea, V. Bertolasi and G. Gilli, *Arzneim. Forsch./Drug Res.*, vol. 36, p. 895, 1986.

[9] A. Schmidt and S. Peroutka, *Mol. Pharmacol.*, vol. 36, p. 505, 1989.

[10] C. Swain, R. Baker, C. Kneen, et al., *J. Med. Chem.*, vol. 34, p. 140, 1991.

[11] M. Hibert, R. Hoffman, R. Miller and A. Carr, *J. Med. Chem.*, vol. 33, p. 1594, 1990.

[12] K. Buchheit, R. Gamse, R. Giger, et al., *J. Med. Chem.*, vol. 38, p. 2326, 1995.

[13] J. George, L. Kier and J. Hoyland, *Mol. Pharmacol.*, vol. 7, p. 328, 1971.

[14] B. Pullman, J. Coubeils, P. Courriere and J. Gervois, *J. Med. Chem.*, vol. 15, p. 17, 1972.

[15] L. Kier, *Pure Appl. Chem.*, vol. 35, p. 509, 1973.

[16] K. Andersen, T. Liljefors, K. Gundertofte, J. Perregaard and K. Bogeso, *J. Med. Chem.*, vol. 37, p. 950, 1994.

[17] A. Albert, Selective Toxicity, London: Chapman and Hall, 1983.

[18] M. Michelson, E. Zeimal, Acetylcholine, New York: Pergamon Press, 1973.

[19] R. Barlow, Introduction to Chemical Pharmacology, London: Academic Press, 1964.

[20] R. Schrage, W. Seemann, J. Klöckner, et al., *Br. J. Pharmacol.*, vol. 169, p. 357, 2013.

[21] J. Saunders, M. Casidy, S. Freedman, E. Harley, L. Iversen, C. Kneen and A. MacLeod, *J. Med. Chem.*, vol. 33, p. 1128, 1990.

[22] S. Kuznetsov, S. Ramsh and A. Zmyvalova, In: *Advances in Science and Technology, Ser. Pharmacology. Chemotherapeutic Drugs*, vol. 25, Moscow, 1991, p. 118 (in Russian).

[23] G. Lambrecht, P. Feifeld, U. Moser, M. Wagner-Roder and L. Choo, *Trends Pharmacol. Sci.*, vol. 10, p. 60, 1989.

[24] R. Feifel, A. Assen, C. Strohmann and R. Tacke, *Naunyn-Schmiedeberg's Arch. Pharmacol.*, vol. 338, p. 193, 1988.

[25] R. Feifel, M. Wagner and C. Strohmann, *Brit. J. Pharmacol.*, vol. 99, p. 445, 1990.

[26] J. Wang, W. Wang, P. Kollman and D. Case, *J. Mol. Graph. Model.*, vol. 25, p. 247, 2006.

[27] E. Pettersen, T. Goddard, C. Huang, G. Couch, D. Greenblatt, E. Meng and T. Ferrin, *J. Comput. Chem.*, vol. 25, p. 1605, 2004.

[28] H. Quastler, The Emergence of Biological Organization, New Haven, London: Yale University Press, 1964.

[29] F. Dukhovich, M. Darkhovskii, E. Gorbatova and V. Kurochkin, Molecular Recognition: Pharmacological Aspects, New York: Nova Science Publishers, 2004, p. 141.

[30] A.A. Jensen, B. Frolund, T. Lijefors and P. Krogsgaard-Larsen, *J. Med. Chem.*, vol. 48, p. 4705, 2005.

[31] M. Volkenstein, General Biophysics, vol. 1, New York: Academic Press, 1983, p. 15.

[32] J. Leysen, C. Niemegeers, J. Van Nueten and P. Laduron, *Mol. Pharmacol.*, vol. 21, p. 301, 1982.

[33] R. Fuller and N. Mason, *Res. Commun. Chem. Pathol. Pharmacol.*, vol. 54, p. 23, 1986.

[34] J. Leysen, F. Awouters, L. Kennis, P. Laduron, J. Vandenberk and P. Janssen, *Life Sci.*, vol. 28, p. 1015, 1981.

[35] D. Burt, I. Creese and S. Snyder, *Mol. Pharmacol.*, vol. 12, p. 800, 1976.

[36] G. Faedda, N. Kula and R. Baldessarini, *Biochem. Pharmacol.*, vol. 38, p. 473, 1989.

[37] L. Toll and S. Snyder, *J. Biol. Chem.*, vol. 257, p. 13593, 1982.

[38] R. Chang, V. Tran and S. Snyder, *Eur. J. Pharmacol.*, vol. 48, p. 463, 1978.

[39] M. Titeler, R. Lyon, K. Davis and R. Glennon, *Biochem. Pharmacol.*, vol. 36, p. 3265, 1987.

[40] W. Billard, V. Ruperto, G. Crosby, L. Iorio and A. Barnett, *Life Sci.*, vol. 35, p. 1885, 1984.

[41] I. Creese, R. Schneider and S. Snyder, *Eur. J. Pharmacol.*, vol. 46, p. 377, 1977.

[42] D. U'Prichard, D. Greenberg and S. Snyder, *Mol. Pharmacol.*, vol. 13, p. 454, 1977.

[43] C. Brown, A. MacKinnon, J. McGrath, M. Spedding and A. Kilpatrick, *Brit. J. Pharmacol.*, vol. 99, p. 803, 1990.

[44] N. Eltayer, G. Kilpatrick, H. Van de Waterbeemd, B. Testa, P. Jenner and C. Marsden, *Eur. J. Med. Chem.*, vol. 23, p. 173, 1988.

[45] B. Syoholm, R. Voutilainen, K. Luomala, J.-M. Savola and M. Scheinin, *Eur. J. Pharmacol.*, vol. 215, p. 109, 1992.

[46] S. Hill and M. Young, *Eur. J. Pharmacol.*, vol. 52, p. 397, 1978.

[47] H. Meltzer and J. Hash, *Pharmacol. Rev.*, vol. 43, p. 587, 1991.

[48] S. Snyder, G. Grenberg and H. Yamamura, *Arch. Gen. Psychiat.*, vol. 31, p. 58, 1974.

[49] J. Leysen, P. Jenner and C. Marsden, *Acta Psychiatr. Scand.*, vol. 69, no. 311, p. 109, 1984.

[50] P. Andersen, F. Gronvald and J. Jensen, *Life Sci.*, vol. 37, p. 1971, 1985.

[51] M. Terai, S. Usuda, I. Kuroiwa, O. Noshiro and H. Maeno, *Japan J. Pharmacol.*, vol. 33, p. 749, 1983.

[52] S. Peroutka and S. Snyder, *Am. J. Psych.*, vol. 137, p. 1518, 1980.

[53] P. Laduron and J. Leysen, *J. Pharm. Pharmacol.*, vol. 30, p. 120, 1978.

[54] H. Meltzer, S. Matsubara and J.-C. Lee, *J. Pharmacol. Exp. Ther.*, vol. 251, p. 238, 1989.

[55] M. Quik, L. Jverssen and A.M.A. Larder, *Nature*, vol. 274, p. 513, 1978.

[56] A. Hancock and C. Marsh, *Mol. Pharmacol.,* vol. 26, p. 439, 1984.

[57] T. Wander, A. Nelson, H. Okazaki and E. Richelson, *Eur. J. Pharmacol.*, vol. 132, p. 115, 1986.

[58] S.-C. Lin, K. Olson, H. Okazaki and E. Richelson, *J. Neurochem.*, vol. 46, p. 274, 1986.

[59] R. Hornung, P. Presek and H. Glossmann, *Naunyn-Schmeideberg's Arch. Pharmacol.*, vol. 308, p. 223, 1979.

[60] S. Snyder and H. Yamamura, *Arch. Gen. Psychiat.*, vol. 34, p. 236, 1977.

[61] A. Giachetti, U. Mittmann and J.E.A. Brown, *Fed. Proceed.*, vol. 46, p. 2523, 1987.

[62] W. Eberlein, G. Trummlitz, W. Engel, G. Schmidt, H. Pelzer and N. Mayer, *J. Med. Chem.*, vol. 30, p. 1378, 1987.

[63] W. Eberlein, W. Engel, G. Trummlitz and G. Schmidt, *J. Med. Chem.*, vol. 31, p. 1169, 1988.

[64] D. Koshland, *Fed. Proc.*, vol. 23, p. 719, 1964.

[65] L. Blumenfeld, Problems of Biological Physics, Berlin: Springer, 1981, p. 120.

[66] D. Boehr, R. Nussinov and P. Wright, *Nat. Chem. Biol.*, vol. 5, p. 789, 2009.

[67] F. Paul and T. Weikl, *PLoS Comput. Biol.*, vol. 16, p. e1005067, 2016.

[68] G. Hammes, Y.-C. Chang and T. Oas, *Proc. Natl. Acad. Sci. USA*, vol. 106, p. 13737, 2009.

[69] H. Grünhagen and J. Changeux, *J. Mol. Biol.*, vol. 106, p. 497, 1976.

[70] H. Grünhagen, M. Iwatsubo and J. Changeux, *J. Biochem.*, vol. 80, p. 225, 1977.

[71] R. Bonner, F. Barrantes and T. Jovin, *Nature,* vol. 263, p. 429, 1976.

[72] F. Barrantes, *Biochem. Biophys. Res. Commun.*, vol. 72, p. 479, 1976.

[73] D. Gakamsky, E. Lewitzki, E. Grell, X. Saulquin, B. Malissen, F. Montero-Julian, M. Bonneville and I. Pecht, *Proc. Natl. Acad. Sci. USA*, vol. 104, p. 16639, 2007.

[74] A. Zurn, U. Zabel, J.-P. Vilardaga, H. Schindelin, M.J. Lohse and C. Hoffmann, *Mol. Pharmacol.*, vol. 75, p. 534, 2009.

[75] K. Solt, J. Johansson and D. Raines, *Biochemistry*, vol. 45, p. 1435, 2006.

[76] J. Schönfelder, D. De Sancho, R. Berkovich, R. Best, V. Muñoz and R. Perez-jimenez, *Commun. Chem.*, vol. 1, p. 59, 2018.

[77] H. Hong, Z. Guo, H. Sun, P. Yu, H. Su, X. Ma and H. Chen, *Commun. Chem.*, vol. 4, p. 156, 2021.

[78] S. Krivov, *J. Phys. Chem. B*, vol. 115, p. 12315, 2011.

[79] F. Mallamace, C. Corsaro, D. Mallamace, et al., *Proc. Natl. Acad. Sci. U.S.A.*, vol. 113, p. 3159, 2016.

[80] C. Mello and D. Barrick, *Proc. Natl. Acad. Sci. U.S.A.*, vol. 101, p. 14102, 2004.

[81] V. Ovchinnikov and M. Karplus, *J. Chem. Phys.*, vol. 140, p. 175103, 2014.

[82] A. Kulshrestha, S. Punnathanam and K.G. Ayappa, *Soft Matter*, vol. 18, pp. 7593–7603, 2022.

[83] E. Eliel, N. Allinger, S. Angyal and G. Morrison, Conformational Analysis , New York: Interscience Publishers, 1965.

[84] L. Graham and R. Diel, *J. Phys. Chem.*, vol. 73, p. 2696, 1969.

[85] P. Krusic and J. Kochi, *J. Am. Chem. Soc.*, vol. 93, p. 846, 1971.

[86] D. Hall and T. Poole, *J. Chem. Soc. (B)*, p. 1034, 1966.

[87] J. Lehn and B. Munsch, *J. Chem. Soc. (D)*, p. 1327, 1969.

[88] E. Eliel, *Accts. Chem. Res.*, vol. 3, p. 1, 1970.

[89] R. Harris and R. Spragg, *J. Chem. Soc. (B)*, p. 684, 1968.

[90] H. Schmid, H. Friebolin, S. Kabuß and R. Mecke, *Spectrochim. Acta*, vol. 22, p. 623, 1966.

[91] R.J. Griffin and R. Coburn, *J. Am. Chem. Soc.*, vol. 89, p. 4638, 1967.

[92] J. Leysen, W. Gommeren, J. Mertens, W. Luyten, P. Pauwels, M. Ewert and P. Seeburg, *Psychopharmacology*, vol. 110, pp. 27–36, 1993.

10 Agonists and Antagonists

10.1 THE PARADOX OUTLINE

In this chapter, we consider neurotransmitters (also named as neuromediators) and mimicking substances that cause profound changes in the conformation of receptors, which lead to opening of ion channels or signaling to other cell subsystems, either directly or indirectly through the participation of GTP-binding G-proteins as a signal transducer. Agonists, such as neurotransmitters, hormones, and synthetic compounds, are substances that can activate the receptor and exhibit specific physiological or neurological effect. Antagonists do not induce receptor activation but instead block receptor signaling. A pharmacological function of antagonist drugs is to protect a receptor from an excessive action of the mediator under pathological conditions. There are also partial agonists activating a receptor, but those agonists cannot, irrespective of concentration growth, elicit the maximum response (efficacy) that is produced by (full) agonists.

It is the paradox that agonists produce a pronounced biological response (in pharmacological terms, have high efficacy) by binding to the receptor with a lower magnitude measurable free energy, when compared with antagonists. Based on the K_d values (Table 10.1), the binding free energy (ΔG) of nicotinic acetylcholine receptor (nAChR) with its antagonist, d-tubocurarine, can be derived as -10.2 kcal/mol, while for nAChR, the binding free energy (ΔG) to acetylcholine (ACh) is -5.8 kcal/mol. Similarly, the ΔG values of N-methylscopolamine (NMS) and ACh binding with the muscarinic acetylcholine receptor (mAChR) comprised -14.0 kcal/mol and -6.7 kcal/mol, respectively. Studies of molecular interactions of agonists and antagonists with receptors have not yet clarified the expressed issue on the mechanism of agonist action.

Paton [13] proposed the hypothesis (usually referenced as "rate theory"), which is widespread nowadays that the type of ligand action (or level of ligand efficacy) on a given receptor depends on the dissociation rate constant of the ligand. Thus, if k_{-1} is high, then the compound would behave as an agonist; if this constant is low, then it would behave as an antagonist; and finally, for intermediate k_{-1} values, the compound would act as a partial agonist. According to Paton, the type of pharmacological effect is dependent on the number of receptors occupied with a given ligand per time unit. In the case of many substances, the relationship between their pharmacological effects and k_{-1} values indeed holds true (e.g., see [14]).

Thus, acetylcholine (ACh) and tetramethylammonium (TMA) ion with k_{-1} values of 10^3 s^{-1} display the agonistic effect on nAChR, whereas hexamethonium and d-tubocurarine with k_{-1} values of $\sim 10^{-1}$ s^{-1} act as antagonists. The same relationship does exist for mAChR ligands.

However, this relationship is not versatile. The lifetime of complexes and affinities to μ-opioid receptors (MORs) of morphine (Figure 7.5) and carfentanyl (Figure 10.1) are higher than those for naloxone, but both morphine and carfentanyl are agonists, whereas naloxone is an antagonist. The

DOI: 10.1201/9781003366669-11

TABLE 10.1
Binding Constants of Some Agonists and Antagonists with Receptors (t Represents Average Complex Lifetime or Ligand's Residence Time). Chemical Structures Depicted in Figure 10.1

	Agonists					Antagonists			
Receptor	Ligand	K_d, M	$k-_1$, c^{-1}	t, c	Ref.	Ligand	K_d, M	$k-_1$, c^{-1}	t, c	Ref.
nAChR	ACh	$5.3 \cdot 10^{-5}$	$1.2 \cdot 10^3$	$8.3 \cdot 10^{-4}$	[1]	TEA	$1.6 \cdot 10^{-3}$	$6.3 \cdot 10^3$	$1.6 \cdot 10^{-4}$	[2]
						d-Tubocurarine	$3.4 \cdot 10^{-8}$	0.17	5.9	[3]
	TMA	$5.2 \cdot 10^{-4}$	$3 \cdot 10^3$	$3.3 \cdot 10^{-4}$	[4]	Hexamethonium	$2.2 \cdot 10^{-8}$	0.48	2	[3]
mAChR	ACh	$1.1 \cdot 10^{-5}$	$1.5 \cdot 10^2$	$6.7 \cdot 10^{-3}$	[5, 6]	TEA	$3.6 \cdot 10^{-4}$	$1.3 \cdot 10^3$	$7.8 \cdot 10^{-4}$	[7]
						Atropine	$1.6 \cdot 10^{-9}$	$7.5 \cdot 10^{-3}$	$1.3 \cdot 10^2$	[8]
						NMS	$5 \cdot 10^{-11}$	$3.3 \cdot 10^{-4}$	$3 \cdot 10^3$	[9]
μ-Opioid	Morphine	$1.5 \cdot 10^{-9}$	$1.6 \cdot 10^{-2}$	62	[10]					
	Carfentanyl	$8 \cdot 10^{-11}$	$1.8 \cdot 10^{-3}$	$5.6 \cdot 10^2$	[12]	Naloxone	$5.2 \cdot 10^{-9}$	$6.2 \cdot 10^{-2}$	16	[11]

behavior of tetraethylammonium (TEA) completely contradicts the Paton hypothesis, according to which TEA is an agonist of cholinergic receptors: actually, this agent displays an antagonistic activity.

Although the molecular weights of antagonists are larger than those of agonists, the exceptions from this rule do exist—the agonist carfentanyl has a larger molecular weight than the antagonist naloxone.

Structural differences between agonistic and antagonistic ligand molecules that are essential for their specific physiological effects cannot always be identified. Thus, the pairs "ACh–tubocurarine" and "TMA–atropine" are examples of agonist–antagonist pairs with distinct structures. Moreover, structurally similar TMA and TEA are antipodes in their action on cholinergic receptors as well as morphine and naloxone, which have many common features in the structure but dissimilar effects as an agonist and an antagonist of MORs, respectively.

To shed some light on the reasons for the functional differences between agonists and antagonists, it is useful to mention that all agonists of the mentioned neuronal receptors are ions in the aqueous medium. As agonists induce conformational changes in the receptors and activation of ion channels, it requires considerable energy. Evidently, the potential energy of ion–ion interaction exceeds in all other types of interactions that are involved in drug–receptor binding (see also Chapter 1). However, antagonists are also ions; therefore, foremost, it may be important to understand the difference between the charged groups of agonists and those of antagonists involved in the interactions with corresponding targets.

It is well known that mediators are separated into two groups based on their specific receptor type. Receptors present in the mammalian central nervous system and neuromuscular synapses are ionotropic (fast acting) and metabotropic (slow acting). The former receptors constitute ion channels, for example, five protein subunits of the nicotinic acetylcholine receptor form an ion channel conducting sodium and potassium ions simultaneously, and two of the subunits serve as acetylcholine receptors. For nAChR, acetylcholine activates the depolarization of the postsynaptic membrane. GABA and glycine receptors are ionotropic inhibitory receptors conducting chlorine ions and keeping membrane potential at a more negative level. Metabotropic receptors are, by itself, not ion channels but regulate ion channels, ion pumps, and other proteins through indirect pathways mediated by G-proteins. mAChR, α- and β-adrenoceptors, serotonin, dopamine, and glutamate receptors as well as neuropeptide receptors are all metabotropic [15].

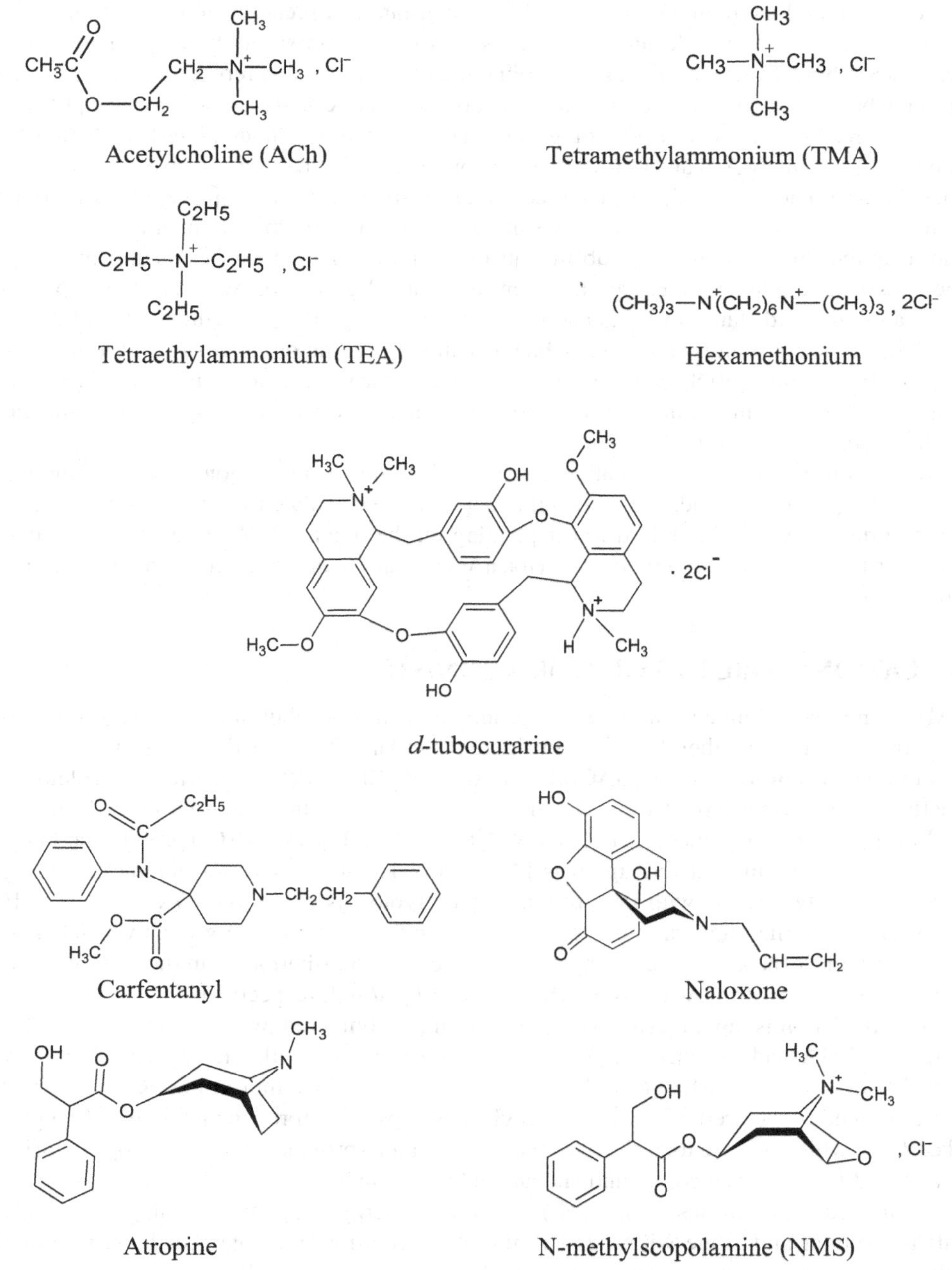

FIGURE 10.1 Chemical structure of agonists and antagonists.

10.2 AGONISTS AND ANTAGONISTS OF CHOLINERGIC RECEPTORS

As mentioned above, a charged group plays a key role in the biological activation of receptors. The substitution of a TMA group for a sterically similar tert-butyl group in the acetylcholine molecule results in a drop in the agonistic activity [16].

However, a liquid medium with a low dielectric constant is a prerequisite for high-energetically favorable electrostatic intermolecular interactions, as well as a proximity between charged groups. The orthosteric binding site in muscarinic cholinergic receptors can be represented as a cup-shaped cavity at the bottom of which the carboxylate anion of the aspartic acid residue is located [17, 18, 19]. The nicotinic receptor has two ACh-binding sites on the $\alpha_1 \gamma$ and $\alpha_2 \delta$ subunit interface, both with two anionic residues, namely Asp and Glu [20]. As the binding pocket has a transverse size that spatially matches the ACh cationic head, water molecules are displaced from the place where reproaching charged groups interact. Within the cavity, van der Waals and electrostatic interaction of aromatic residues near the anionic group with substituents on the nitrogen atom reaches its maximum.

The electrostatic interaction energy between the charged groups of ACh and its receptor counterpart at a minimum distance of 3.3 Å can be estimated through the following method. Setting that the local dielectric constant is equal to 3, which is within the common range of 2–4 for proteins [21, 22, 23], the E_{max} is thus −30.8 kcal/mol. Energy expenses for water displacement would reduce (by absolute value) the maximum interaction energy E_{max}, and the opposite change (toward increasing) occurs because of the binding of the methyl groups.

However, contribution from the cationic groups of ACh and other agonists to the binding free energy under equilibrium conditions is much lesser than the calculated value (see Table 10.2).

The reported K_d values slightly differ depending on the organism. We refer mainly to the data obtained for frog rectus abdominis muscle (for nAChR) and mammalian intestinal smooth muscle (for mAChR).

10.3   CATIONIC GROUPS OF AChR AGONISTS

As TMA completely imitates the cationic group in ACh, contributions of these groups to the binding free energy can therefore be considered similar. The contribution from the cationic group in molecules of agonists of mAChR from the $CH_3(CH_2)_n N^+(CH_3)_3$ series was calculated by subtracting the contributions of the CH_2- and CH_3-groups and uncharged moieties from the ΔG value. In this set, the substances $CH_3(CH_2)_3 N^+(CH_3)_3$ and $CH_3(CH_2)_4 N^+(CH_3)_3$ with the butyl and pentyl groups are the most active agonists [24]. The calculated dispersion interaction energy of the butyl and pentyl groups with the protein surface groups is equal to −1.7 and −2.1 kcal/mol, respectively, where the interaction energy for the methyl and methylene groups was taken as 0.5 and 0.4 kcal/mol, respectively (see Chapter 1). Hence, the contribution from the cationic group to the binding enthalpy with mAChR is −3.18 and −3.23 kcal/mol, respectively.

Such a calculation is quite reasonable as seen from the compared ΔG values of −3.13 and −7.18 kcal/mol for TMA and decyltrimethylammonium, respectively, with the experimental K_d value equal to $5.0 \cdot 10^{-3}$ M and $5.3 \cdot 10^{-6}$ M [24]. A difference of 4.05 kcal/mol in the ΔG values between them is attributed to the presence of nine methylene groups; therefore, one methylene group shares 0.45 kcal/mol, which is close to derived from the calculation (provided that the entropy contribution is not changed for a charged compound and an uncharged analog).

The estimated contributions of $-N^+(CH_3)_3$ to the free energy of agonists binding to mAChR and nAChR are similar and range within 3–4 kcal/mol. In addition, we obtained the same estimated contributions for the cationic group of two partial agonists as shown in Table 10.2.

10.4   CATIONIC GROUPS OF AChR ANTAGONISTS

It is most amazing that antagonists do not differ from agonists in terms of the binding free energy of their cationic groups to cholinergic receptors. The electrostatic interaction energy can be estimated through substitution of the charged nitrogen atom for the neutral carbon atom, as entropy contribution remains the same in this case. In such a manner, the neutralization of the quaternary nitrogen group in potent atropine-like compound benziloylcholine (Figure 10.2) by producing

TABLE 10.2
K_d Values of Ligand–Cholinergic Receptor Complex Formation and the Contribution ($\Delta\Delta G$) of the Cationic Group

Ligand	Receptor	K_d, M	Ref.	$-\Delta G$, kcal/mol	Contribution of Cationic Group, $\Delta\Delta G$, kcal/mol
		Agonists			
TMA	nAChR	$5.2 \cdot 10^{-4}$	[4]	4.46	4.46
TMA	mAChR	$5.0 \cdot 10^{-3}$	[24]	3.13	3.13
ACh	nAChR	$1.4 \cdot 10^{-6}$	[25]	7.96	4.46*
ACh	mAChR	$7 \cdot 10^{-6}$	[26]	7.02	3.13*
$CH_3(CH_2)_3N^+(CH_3)_3$	mAChR	$2.6 \cdot 10^{-4}$	[24]	4.88	3.18**
$CH_3(CH_2)_4N^+(CH_3)_3$	mAChR	$1.2 \cdot 10^{-4}$	[24]	5.33	3.23**
		Partial agonists			
$CH_3(CH_2)_6N^+(CH_3)_3$	mAChR	$2.4 \cdot 10^{-5}$	[24]	6.28	3.08**
$CH_3(CH_2)_7N^+(CH_3)_3$	mAChR	$1.6 \cdot 10^{-5}$	[24]	6.53	3.08**
		Antagonists			
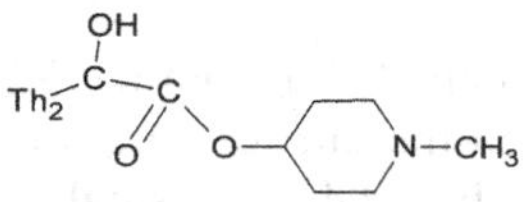	mAChR	$5.6 \cdot 10^{-11}$	[27]	14.60	2.75**
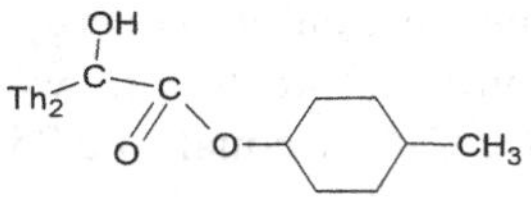	mAChR	$5.8 \cdot 10^{-10}$		13.10	
$CH_3(CH_2)_9N^+(CH_3)_3$	mAChR	$5.3 \cdot 10^{-6}$	[24]	7.18	3.08**
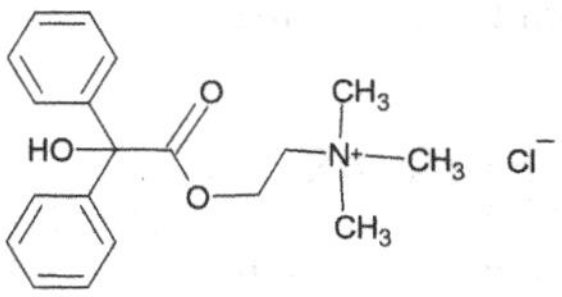	mAChR	$2 \cdot 10^{-9}$	[28]	11.8	3.17
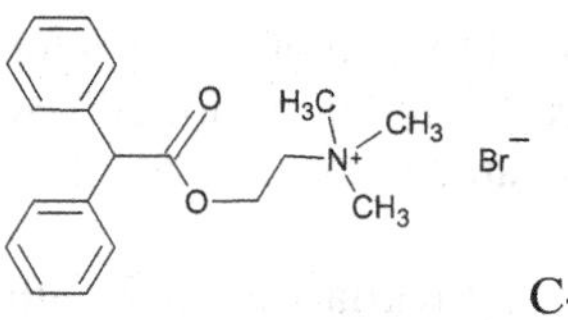 C-10.16	mAChR	$4.7 \cdot 10^{-8}$	[27]	10.4	—
TEA	mAChR	$3.6 \cdot 10^{-4}$	[7]	4.68	3.10**
TEA	nAChR	$1.6 \cdot 10^{-3}$	[2]	3.80	3.10**

Designations: Th - 2-thienyl
 * By analogy with TMA
** Calculated values

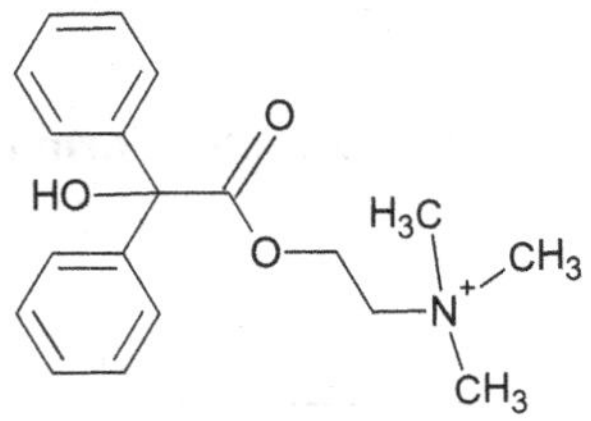

FIGURE 10.2   Chemical structure of benziloylcholine.

benziloyldimethylbutanol causes a 3.1-fold decrease in the affinity [28, 29], corresponding to $\Delta\Delta G = 0.67$ kcal/mol.

The substitution of a $-C(CH_3)_3$ group for the methyl group in benziloyldimethylbutanol makes it possible to estimate the contribution from three methyl groups: the affinity decreases by 69.6 times, and the overall $\Delta\Delta G$ becomes 2.50 kcal/mol or 0.83 kcal/mol per one methyl group. A direct change from $-N^+(CH_3)_3$ for the methyl group in benziloylcholine results in a 217-fold reduction in the affinity to mAChR [28], corresponding to a 3.17 kcal/mol decrease. With the subtraction of the electrostatic part (-0.67 kcal/mol), we again derive the value of -2.5 kcal/mol, the contribution of three methyl groups, or -0.83 kcal/mol per each of them. Therefore, the experimental value of -3.17 kcal/mol is indeed a partial contribution of the benziloylcholine cationic group to the free energy of complex formation with mAChR.

The binding free energy of the cationic group to mAChR can also be calculated from the K_d data for substance C-4.11 (Figure 7.9), and its uncharged analog in which the nitrogen atom is substituted with the carbon atom. The affinity of the uncharged analog is by 10.4 times smaller than that of C-4.11 or by 1.45 kcal/mol, and it can be attributed to changes in Coulomb energy. Having the estimated contribution of alkyl radicals in the cationic group (1.3 kcal/mol), we obtained the binding free energy value for the cationic group of substance C-4.11 equal to 2.75 kcal/mol.

The contribution of the TEA cationic group is estimated with subtracting the contribution of four methyl groups because of dispersion interaction energy (-2 kcal/mol) from the ΔG value.

10.5   COMPARISON OF AGONISTS AND ANTAGONISTS OF mAChR

Equal contributions of the cationic group in agonists, partial agonists, and antagonists to the binding free energy of the cholinergic receptors led to a conclusion that the equilibrium methods for K_d determination may be not suitable to characterize high-energy stages during agonist–receptor interactions due to their fast rates. The methods for K_d measurement are pertaining to long-lived states with low-energy interactions. Notwithstanding this, the equilibrium parameters provide information to distinguish between agonists and antagonists.

An agonist's function is gradually changed to the role of antagonists and partial agonists with increasing contribution of polar groups and nonpolar aromatic moieties at the acyl end of the ACh-like molecule to the energy of interaction with cholinergic receptors. Particularly, benziloylcholine (Figure 10.2), containing the same cationic group as that in ACh, exerts the antagonistic effect on muscarinic cholinergic receptors.

Based on Table 10.2 data, the free energy threshold Q to distinguish muscarinic cholinergic receptor agonists and antagonists can be proposed, lying between the most potent agonist and the least potent antagonist in a series of compounds with a similar scaffold. For example, $\Delta G_{ACh} > Q > \Delta G_{C-10.16}$, or $-7.0 > Q > -10.4$ (in kcal/mol). Hence, Q variation is no more than 3.4 kcal/mol.

The Q value can also be calculated for a series of alkyltrimethylammonium ions $CH_3(CH_2)_nN^+$ $(CH_3)_3$ [24]. For the most potent agonist among these ions $CH_3(CH_2)_3N^+(CH_3)_3$, its ΔG value is -4.9

kcal/mol, while for the antagonist decyltrimethylammonium $CH_3(CH_2)_9N^+(CH_3)_3$, ΔG is -7.2 kcal/mol. It gives a Q variation of 2.3 kcal/mol.

A value of 2.8 kcal/mol might be proposed as an average affinity difference between the agonists and antagonists of muscarinic cholinergic receptors. The cationic group and the remaining uncharged groups of a ligand molecule form a stable (long-lived) position on the receptor in an opposite manner: the cationic group pulls a ligand toward rapprochement with the receptor's anionic group, whereas the uncharged moiety retains from this.

It has long been known that the change from agonistic to antagonistic activity is associated with increasing molecular mass of ligand molecules. Here we have evaluated only the free energy threshold between agonists and antagonists.

The suggested difference in the interaction with muscarinic choline receptors between agonists and antagonists [30] is illustrated in Figure 10.3. The anionic site of the receptor is placed in a cavity.

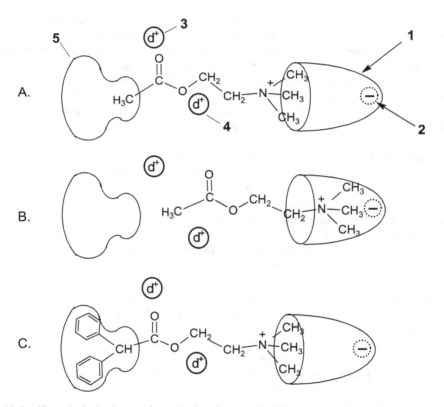

FIGURE 10.3 Hypothetical scheme of agonists' and antagonists' interaction with mAChR, as an example of ACh and compound C-10.16.

Designations:

1 –receptor anionic site in the cavity; 2 –anionic residue; 3.4 – binding sites of acetylcholine oxygen atoms; 5 – hydrophobic area of the receptor.

A – inactive position of ACh, B – active position of ACh; C – stable position of compound 10.16.

The hydrophobic region around the neck of the receptor site and water molecules inside the cavity are not depicted.

It follows from the model of Figure 10.3 that an increase in the cationic group size, preventing charged groups of the receptor and ligand from being in close position, might be a main factor that differentiates agonists from antagonists. TMA is an agonist, while TEA with a bigger cationic head is an antagonist despite the low strength of its complex with muscarinic cholinergic receptors, and its k_{-1} value is comparable to that of ACh and TMA.

Based on the equilibrium parameters of antagonist binding, it is possible to determine the depth of the orthosteric receptor pocket with the anionic group. The estimated Coulomb energy for an interaction between one methyl group in the cationic head of substances C-4.11 and benziloylcholine and the receptor's anionic group is 1.1 kcal/mol given that entropy contribution is negligible. Substituting this energy value and the formula for the local dielectric constant (1.35) into Equation (1.6), the distance between the charge centers is calculated as 8 Å. The ionized groups interact at this distance keeping their monomolecular water shells.

10.6   COMPARISON OF AGONISTS AND ANTAGONISTS OF nAChR

Returning to the basic difference between agonists and antagonists in their interaction with nicotinic cholinergic receptors, it is likely that high-energy stages of agonists' action are hidden primarily because of their fast transience. Electrophysiological experiments with measurement time of the current impulse transmission across the synapse have shown that the mean lifetime of the channel opening, which is activated by ACh through its effect on nicotinic cholinoreceptor, is 1 ms [31, 15], whereas a complex of ACh interacting with nAChR has an average lifetime of <1 ms [1, 32]. The K_d constants pertain to long-lived stages, such as receptor desensitization that are the least energetically favorable. Thus, the rapid reversibility of receptor activation after its high-energy binding with an agonist must be explained.

This paradox can be explained using a hypothetical scheme [30] of fast stages during ACh interaction with the orthosteric site of the nicotinic cholinergic receptor depicted in Figure 10.4.

I   The stage when the receptor is inactive and the ion channel is closed.

II The stage when the ligand begins approaching the residues in the active site, conformational changes start, and water molecules are displaced from the interaction region. The ion channel starts to open. The extent of receptor group draft into the cavity volume and the cavity opening reflects the depth of conformational changes in the receptor.

III The stage of the strongest interaction between the ligand and the receptor groups, when the cationic head and receptor groups are in the closest contact. Water molecules are displaced from the contact site; therefore, the local dielectric constant is low. The receptor acts as a mechanical machine that opens the ion channel.

IV As the cavity further opens, the cationic group of agonists is no longer a barrier for water molecules. Hydration shell electrostatic screening causes an increase in the dielectric constant and a reduction in the Coulomb ion–ion interaction energy, and the agonist–receptor complex dissociates within a very short time.

Breakdown of the complex returns the receptor to the inactive stage I.

The qualitative model shown in Figure 10.4 can be compared to the more elaborated Auerbach "catch-and-hold" model for agonist activation of the nicotinic receptor [33, 34]. In the Auerbach model, the closed (C) and open (O) states of the receptor have different degrees of affinity to the agonist. The agonist-forming complex AC ("catch") undergoes a series of fast conformational transitions ("hold"). The agonist-bound state transforms AC by lowering the free energy barrier for transition to the agonist complex with open state AO ("zipper" mechanism). Notably, the association rate of ACh-like compounds with the O state is controlled by diffusion [35]. This is shown in phase IV of the model depicted in Figure 10.4, where the cavity is easily accessible to both water and

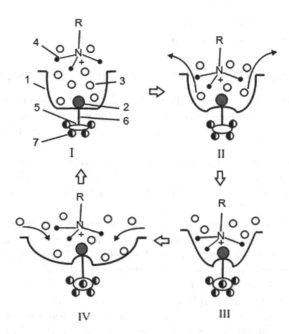

FIGURE 10.4 Hypothetical scheme of binding stages between agonists of common formulae R-N$^+$(CH$_3$)$_3$ with the nicotinic cholinergic receptor's orthosteric site.

Designations:

(1) receptor's site; (2) receptor's binding groups; (3) water molecules; (4) methyl group; (5) an orifice of the ion channel[2]; (6) receptor-channel coupling entity; (7) metal cations responsible for transmembrane current.

agonist molecules, provided the cavity in the isolated O state is initially empty. The Auerbach model also proposes that there are hidden short-lived fast transformations of the initial agonist complex that can be barely detected.

For neuromuscular synapses (skeletal muscles), ACh removal from the synapses rapidly (approximately 1 ms) to take the next signal is a feature. This is facilitated by ACh diffusion from the synaptic cleft and its reverse capture from the pre-synaptic terminal, but the main role is attributed to acetylcholinesterase. Because of the high rate of cleavage of free (unbound) ACh by acetylcholinesterase (one molecule per 0.1 ms) and the high content of AChE in the postsynaptic membrane, within approximately 0.1 ms after the ACh enters the synaptic cleft, the concentration of this mediator in the gap becomes so low that nAChR activation becomes insignificant [15]. As a result, the equilibrium shifts toward the dissociation of the ACh–nAChR complex. This mechanism of nAChR reactivation seems logical, if not for one circumstance. The undoubted existence of high-energy stages of interaction between ACh and nAChR should have caused a long lifetime of the complex (see Chapter 6), which is not observed.

We need to consider the two types of conformational changes separately. The first conformational change is the mutual adjustment of the ligand and the receptor structure during complex formation. This type of conformational change is characteristic for both agonists and antagonists. The second conformational change refers to conformational changes in the receptor that determine the operation of the ion channel. This type of change is associated only with the effect of agonists. The fact that these are different types of conformational changes can be seen in the example of a TMA agonist, which causes changes only in the second type. As expected, the work of the receptor-channel junction is much more complicated than that shown in Figure 10.4 [15]. One should consider the potential dependence of the conformational structure of the complex on depolarization or hyperpolarization as well as on the operation of ion pumps that restore the resting potential of the cell.

10.7 ROLE OF CHARGED GROUPS IN AGONISTS AND ANTAGONISTS

Returning to the question of the functional distinction of charged groups in agonists and antagonists, these groups are necessary for antagonists to recognize the cholinergic receptor. However, charged agonist groups, in addition to recognition, are required to change the conformation of nAChR, activating ion channels.

All acetylcholine-like activities are sharply diminished with successive substitutions of methyl groups for hydrogen atoms in the cationic group of agonist molecules [16]. In physiological experiments with acetylcholine and its derivative N,N-dimethylaminoethylacetate (e.g., $-N^+H(CH_3)_2$ group is substituted for $-N^+(CH_3)_3$), muscarinic cholinergic activity (cat intestine tissues) decreases by 412-fold, while nicotinic cholinergic activity (frog abdomen straight muscle tissues) decreases by 600-fold [36]. Such a strong inactivation is not explicable by eliminating the contribution from one methyl group to the binding energy. Substitution of a $-N^+(CH_3)_3$ group for an $-N^+H_3$ group in the ACh molecule results in almost complete abolishment of the muscarinic agonist activity [24]. According to the proposed schemes in Figure 10.3 and Figure 10.4, the dropped agonist activity might be caused by the incomplete displacement of water molecules from the receptor anionic site because of a decreased size of the cationic group. In addition, a water molecule binds to the $-N^+-H$ group much stronger than that to the $-N^+-CH_3$ group that can impede the close contact between the ligand and receptor groups. However, the presence of the tertiary nitrogen atom in cyclic moieties is essential for a profound ACh-like activity posed by some compounds, such as nicotine (Figure 10.5) and arecoline (Figure 10.6) in the protonated form, the agonists of nicotinic and muscarinic ACh receptors, respectively.

With superposition of oxygen and nitrogen atoms in arecoline and ACh, which constitute the pharmacophore of mAChR ligands (see Chapter 9), the groups marked with * in arecoline are in the same position as that of the methyl groups in the ACh cationic group. It is probable that the groups * and the protonated nitrogen atom in nicotine constitute the cationic group. In this case, the cationic groups in both nicotine and arecoline have the same size as that in the acetylcholine molecule. To gain insight into the high activity of protonated cyclic amines, the hydrophilic $^+N–H$ bond may be directed outside the receptor's group so that complex formation is not prevented. Such direction of the N–H bond in the nicotine molecule is indirectly evidenced from its unchanged agonistic effect after methylation (quaternization) of the nitrogen atom in the pyrrolidine cycle [37].

A similar binding affinity change in agonists with a modified cationic group for both muscarinic and nicotinic cholinergic receptors suggests a similar structure in the binding site. Nevertheless, responses that are mediated by muscarinic cholinoreceptors for different neurons are slower by 10–1000 times than those with nicotinic receptors [38]. Signals from the central nervous system to skeletal muscles are transmitted through nicotinic receptors. A typically rapid response that is mediated through nAChR is clearly essential for animals to survive. Muscarinic cholinergic receptors mediate the signal transmission to smooth muscles of vessels and internal organs that do not require a quick response. The muscarinic receptor is not directly coupled with the ion channel [39], which might

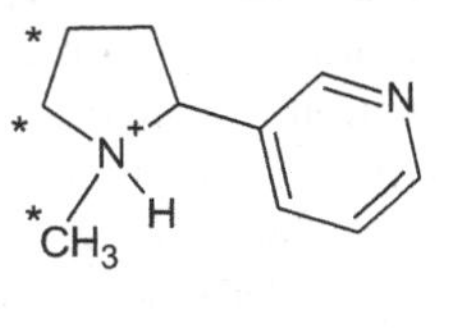

FIGURE 10.5   Structure of nicotine (protonated).

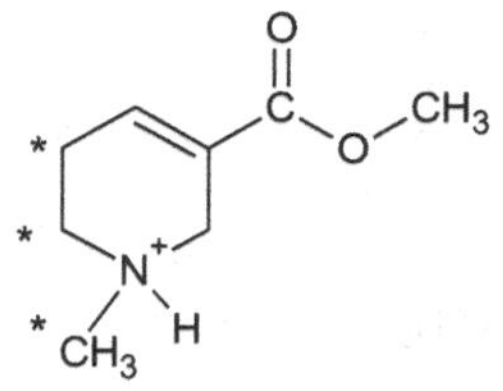

FIGURE 10.6   Structure of arecoline (protonated).

explain the slow rate of the response. The latent response of mAChR may be mediated by slow stages following likely fast binding of the agonist to mAChR. It is evident that the formulated paradox needs further research.

10.8   PARTIAL AGONISTS OF mAChR

Partial agonists of the muscarinic acetylcholine receptors have not been considered yet. These agonists are ligands that cannot produce the maximum stimulus of the receptor (in contrast to acetylcholine and other full agonists) irrespective of the increasing concentration.

The moderate activity of partial agonists is rationalized with their intermediate affinity, ranging from complete agonists to antagonists. The fact that partial agonists are unable to elicit the maximum contracture in muscle tissue samples is explained by their weak intrinsic activity (efficacy) on the corresponding receptor [24, 40]. It is unclear what stops on a halfway the approaching trimethylammonium groups of partial agonists listed in Table 10.2 to the muscarinic receptor's anionic group, thus eliminating the typical ACh-like activity.

Recently, partial agonists of mAChR with bitopic properties (both bound orthosteric and allosteric sites) were discovered, such as McN-A-343, and some benzyl quinolone carboxylic acid (BQCA) covalently linked by the $-(CH_2)_n-$ chain to agonist fragments (iperoxo, acetylcholine, oxotremorine, and TMA) [41, 42, 43]. Thus, the allosteric site of mAChR might also participate in the regulation of partial agonism of a certain compound.

The following explanation for partial agonists' mode of action can be proposed: the incomplete response from a muscle tissue is due to an insufficient number of activated receptors, irrespective of the partial agonist concentration. Why is it possible? The relationship of affinity for mAChR agonists, partial agonists, and antagonists allows the estimation of the discriminant Q (affinity threshold) between agonists and partial agonists of mAChR, on the one hand, and partial agonists and antagonists, on the other hand. Thus, the energy discriminant is very fragile, being less than 1.5 kcal/mol (see Table 10.2), which is close to the thermal motion energy of molecules under

TABLE 10.3
Affinity (K_d) of Some Agonists and Antagonists to MOR

Ligand	K_d, nM	$-\Delta G$, kcal/mol	Ref.
Morphine	1.17	12.1	[44]
Fentanyl	1.35	12.1	[44]
Alfentanyl	7.39	11.1	[44]
Sufentanyl	0.14	13.4	[44]
Carfentanyl	0.024	14.4	[26]
Naloxone	5.2	11.3	[11]
Naltrexone	0.77	12.4	[45]

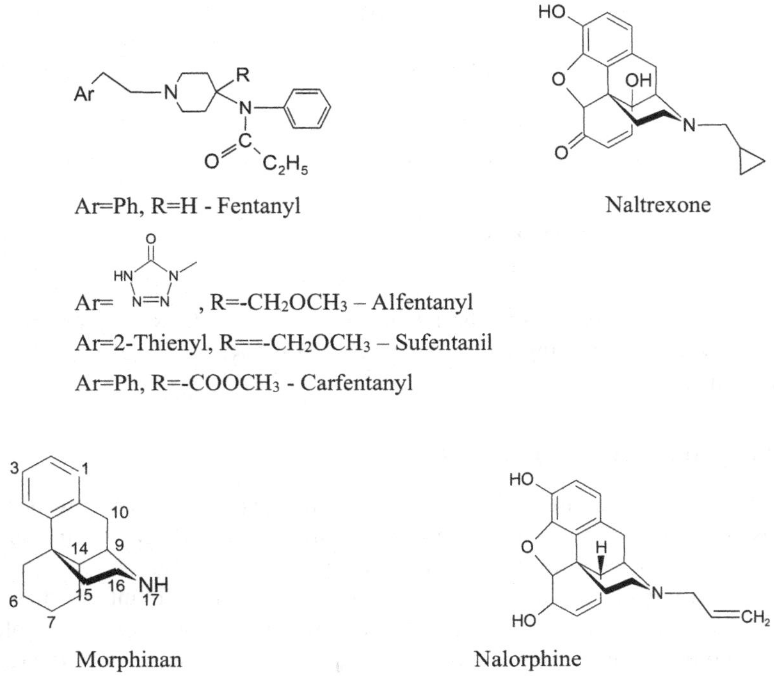

Ar=Ph, R=H - Fentanyl

Ar= (triazolone), R=-CH$_2$OCH$_3$ – Alfentanyl

Ar=2-Thienyl, R==-CH$_2$OCH$_3$ – Sufentanil

Ar=Ph, R=-COOCH$_3$ - Carfentanyl

Naltrexone

Morphinan

Nalorphine

FIGURE 10.7   Structure of some opioid agonists and antagonists.

physiological conditions (330 K, 0.9 kcal/mol). In other words, molecules of partial agonists might balance on a verge: to produce receptor activation such as ACh or to produce receptor blocking similar to atropine, getting into one or another free energy well. The ratio between molecules of partial agonists producing activation or blocking is determined by statistical distribution. Thus, partial agonists display the distinct effect because of a low proportion of molecules to produce a complete ACh-like response, not to their intermediate position in the receptor binding pocket. An increase in the concentration of partial agonists does not produce the maximum response because of an increasing blocking effect.

10.9   AGONISTS AND ANTAGONISTS OF μ-OPIOID RECEPTOR

The MORs mediate the analgesic and narcotic effects of agonists. In contrast to cholinergic receptors, the most active opioid agonists have greater affinities than antagonists, as can be seen for fentanyl analogs (Table 10.3). However, the affinity of naltrexone (Figure 10.7), the antagonistic agent, is higher than that of agonists, such as fentanyl, alfentanil (Figure 10.7), and morphine (Figure 7.5). Even if agonists have more affinity than antagonists, a difference in the complex formation energies is too little to explain their effects on MOR.

Morphinan-alike agonists and antagonists have no clear structural differences. Thus, nalorphine, such as naloxone, contains an N-allyl group but is not a pure antagonist. As seen from a comparison of nalorphine (Figure 10.7) and naloxone (Figure 10.1) structures, a balance between agonistic and antagonistic activities is conditional upon the substituents at position 6 in morphinan, the presence or absence of a double bond at position 6 or of the OH-group at position 14. However, the main role in the appearance of agonistic/antagonistic effects is played by the cationic group, as evidenced from the fact that the allyl-to-methyl substitution in naloxone yields hydroxymorphine that behaves as an agonist [46].

TABLE 10.4

Affinity Constants of Opioid Ligands and Their N-Methylated Analogs

Type of Activity	Ligand	Ref.	K_d, nM	K_{d1}/K_{d2}	$\Delta\Delta G$, kcal/mol
Antagonist	Naloxone	[49]	1.8	10.1	1.37
	N-methyl-naloxone		18.2		
Antagonist	Naltrexone	[49]	1.08	7.4	1.18
	N-methyl-naltrexone		8.0		
Agonist–antagonists	Levallorphan	[50]	1.6	7.8	1.21
	N-methyl-levallorphan		12.5		
Agonist–antagonists	Nalorphine	[50]	3.3	15.2	1.61
	N-methyl-nalorphine		50		
Agonist	Fentanyl	[26]	1.2	15000	5.68
	N-methyl-fentanyl		18000		
Agonist	Carfentanyl	[26]	0.024	1750	4.41
	N-methyl-carfentanyl		42		
Agonist	Morphine	[51]	88	20.7	1.79
	N-methyl-morphine		1823		

Protonated nitrogen atom substituents with a larger size than the methyl group may likely hamper the approaching ligand to receptor's charged groups when binding to the nonpolar residues of the receptor binding pocket, thus imparting properties of agonist–antagonists or full antagonists. Morphine-like analgesics with the N-methyl group have no such hampering, as well as normorphine, an agonist, with $K_d = 5.3 \times 10^{-9}$ M [9]. Interestingly, the polar His297 residue attributed to the MOR binding cavity is important for distinguishing agonists and antagonists [47]. The mutation H297A affects naloxone to produce agonist-like activity but decreases its binding affinity [48]. It is supposed that His297 forms a hydrogen bond with the OH-group in position 3 of morphinan-like compounds. Based on this, we may guess that its disruption in mutant MOR produces better opportunity for the charged nitrogen group to be close to the binding pocket's anionic group.

The major role in the manifestation of the agonistic activity of fentanyl and its analogs is played by the greater conformational mobility of the piperidyl group itself than its mobility in morphinan derivatives. It is plausible that fentanyl and its analogs are very active agonists because of the close contact of their charged groups with the receptors when compared with those of morphine. Aromatic groups connected by a flexible ethyl chain seemingly do not prevent the mobility of the charged nitrogen atom.

In fact, neither the affinity level nor the structural peculiarities enable to fully discern the pharmacological effects of MOR ligands. Another conclusion can be drawn for ligands and their N-methylated quaternary analogs (Table 10.4): we see the reduction in affinity upon quaternization, which is minimal for antagonists, maximal for agonists (except morphine), and intermediate for agonists–antagonists[1].

With a transition from fentanyl and carfentanyl to their N-methylated derivatives, the affinity decreases, respectively, by 15,000 and 1750 times, equivalently to a change in the binding free energy ($\Delta\Delta G$) of 5.7 and 4.4 kcal/mol (on average ~5 kcal/mol). The significant difference between the tertiary and quaternary analogs might be attributed to the fact that an additional N-methyl group prevents the ligand's cationic head from approaching the area of the receptor's anionic residues, as for the agonist TMA/antagonist TEA interactions with cholinergic receptors. With the inner repulsion between N-methylated compounds and an additional methyl group, the contribution of fentanyl and carfentanyl cationic groups into $\Delta\Delta G$ can be estimated as 6–7 kcal/mol.

Such relatively high free energy to be required for interactions between the charged groups of these agonists and the receptor can be explainable for the fast stages hidden from the equilibrium experiment, such as ACh binding stages to nicotinic cholinergic receptors. MORs participate in a relatively slow signaling pathway (compared with nicotinic cholinergic receptors) [15, 52], and high-energy stages are longer, as confirmed by experimental data obtained under equilibrium conditions. However, it is unlikely that 6–7 kcal/mol can be totally attributed to the high-energy stage. Perhaps, this case indicates the major part of the hidden free energy changes. Assuming an equilibrium distance between the ligand and receptor group charge centers of 4–5 Å and proportioning the resultant agonist interaction energy to the Coulomb energy, we can evaluate the local dielectric constant in the region of the ion–ion (salt-bridge) interaction. Equation (1.35) gives the calculated ε value of 10–11, but the actual ε value is smaller accounting for estimated free energy changes only. A low dielectric constant indicates that water molecules are completely or largely displaced from the MOR binding pocket with the anionic group during agonist binding.

Thus, energy changes because of quaternization regularly decrease in a row: agonists > agonists–antagonists > antagonists.

Pert and Snyder [53] found that the affinity of opiate ligands (with agonistic, mixed agonistic–antagonistic, and antagonistic types) changes in a distinct manner in the presence and absence of NaCl in an aqueous medium (Table 10.5).

Based on the data, the authors hypothesized the existence of opioid receptors in two conformations, agonist-favored and antagonist-favored, that changes in a process of mutual transformation. Therefore, conformational changes in opioid receptors under the effect of ions underlie the basic mechanism involved in the molecular recognition of specific ligands.

It is not unique that MOR ligands change their affinity under the effect of inorganic ions, and this has been known for ligands of mAChR, glycine, serotonin, and other receptors. In particular, the affinity of carbachol, the agonist of muscarinic cholinergic receptors, decreases by 280 times in the presence of Na^+ ions across a concentration range of 0.1–1 M [54]. Na^+, K^+, and Li^+ ions significantly reduce the affinity of 5-hydroxytryptamine (serotonin) to 5-HT receptors [55].

There might be another cause of the change in the MOR ligand affinity in the presence of NaCl. Dropping affinity is associated with higher energy costs for the displacement of hydrated Na^+ and Cl^- ions from the binding cavity, than the replacement of water molecules only. The affinity decreases as the energy of interaction between the ligand and MOR charged groups declines in the following

TABLE 10.5
Effect of NaCl solution (100 mM) on Ligand's Affinity to MOR of the Rat Brain Homogenate as Determined by Competition with ³H-Naloxone (Data Extracted from [53])

Type	Ligand	K_d, nM NaCl (A)	K_d, nM No NaCl (B)	Ratio A/B	$\Delta\Delta G$, kcal/mol
Antagonists	Naloxone	1.5	1.5	1	0
	Naltrexone	0.5	0.5	1	0
Agonist-antagonist	Nalorphine	4.0	1.5	2.67	0.58
	Pentazocine	50	15	3.33	0.71
	Cyclazocine	1.5	0.9	1.67	0.30
	Levallorphan	2.0	1.0	2	0.41
Agonists	Morphine	110	3.0	36.7	2.13
	Normorphine	700	15	46.7	2.27
	Methadone	200	7.0	28.6	1.98
	Oxymorphone	30	1.0	30	2.01

order: "agonists" > "antagonists–agonists" > "antagonists." The presence of NaCl does not affect the affinity of naloxone and naltrexone, as their cationic groups cannot enter the binding cavity of the anionic residue and therefore are not involved in the displacement of Na^+ and Cl^- ions. The competition of the ligand with inorganic ions depends on the ion charge, the radius of the hydrated ion radius, and the number of displaced ions. Possible changes in the receptor conformation under the effect of NaCl cannot be excluded; however, they are rather not physiologically significant.

Special attention should be paid to the similar order of changes in the affinity from agonists to antagonists (Tables 10.4 and 10.5). Although the data in these tables were obtained independently, they show a common pattern: the consistent decline in the affinity reflects increasing difficulties for larger ligand cationic groups to approach the anionic receptor group. As for ligands mentioned in Table 10.4, steric hindrances or competition with Na^+ and Cl^- ions might cause such difficulties as in the cases listed in Table 10.5.

10.10 CONCLUSION

1. The paradox of cholinergic receptor agonists, which is represented by the discrepancy between high efficacy and low binding energies, is explained by the fact that the high-energy agonist–receptor binding stages are hidden because of their high rate route and high reversibility.
2. The anionic site of receptors that participate in neurotransmission plays a major role in conformational changes in receptors under the effect of agonists.
3. The equilibrium constant K_d pertains to long-lived stages and does not reflect receptor activation by agonists.
4. The rapid dissociation of a complex between ACh and its receptor might be caused by a decrease in the local dielectric constant during the interaction of their charged groups, not only by cholinesterase action as is commonly accepted.
5. It is possible to reveal a high-energy stage of ion–ion interaction for agonists of slowly acting MORs.
6. The hypothesis on the existence of two MOR conformations suitable for binding with either agonists or antagonists is not necessary. Different effects of the NaCl solution on the affinity of agonists and antagonists can be explained by competition of the ligands with Na^+ and Cl^- ions compared with water molecules for binding to the receptor's anionic group.
7. A relationship of the agonistic and antagonistic properties with the rate constant k_{-1} of the ligand–receptor complex according to Paton is nominal: it occurs for some substances and receptors and is absent for the others.

NOTES

1 A term "agonist-antagonist" is used for opioid receptor ligands, instead of partial agonists.
2 Separation of the ion channel and the receptor shown in Figure 10.4 is only for illustrative purpose, as, in reality, the channel is folded by subunits of the receptor [15].

REFERENCES

[1] V. Skok, A. Selyanko and V. Derkach, Neuronal acetylcholine receptors, Moscow: Nauka, 1987, p. 102 (in Russian).
[2] G. Akk and H. Steinbach, *J. Physiol.*, vol. 551, p. 155, 2003.
[3] P. Asher, W. Large and H. Rang, *J. Physiol. (Gr. Brit.)*, vol. 295, pp. 139–170, 1979.
[4] Y. Zhang, J. Chen and A. Auerbach, *J. Physiol.*, vol. 486, p. 189, 1995.
[5] M. Watson, H. Yamamura and W. Roeske, *J. Pharmacol. Exp. Ther.*, vol. 237, p. 419, 1986.
[6] B. Sakmann, A. Noma and W. Trautwein, *Nature*, vol. 303, p. 250, 1983.
[7] M. Caulfield, *Neurosci. Lett.*, vol. 127, p. 165, 1991.

[8] I. Kloog and M. Sokolovsky, *Biochem. Biophys. Res. Commun.*, vol. 81, p. 710, 1978.

[9] M. Waelbroeck, M. Tastenoy, J. Camus and J. Christophe, *Mol. Pharmacol.*, vol. 38, p. 267, 1990.

[10] S. Zaitsev, K. Yarygin and S. Warfolomeev, Drug addiction. Neuropeptide morphine receptors, Moscow: Moscow State University Press, 1993, p. 88 (in Russian).

[11] J. Leysen, W. Gommeren and C. Niemegeers, *Eur. J. Pharmacol.,* vol. 87, p. 209, 1983.

[12] M. Titeler, R. Lyon, M. Kuhar, J. Frost, R. Dannals and S. Leonhardt, *Eur.J.Pharmacol.*, vol. 167, p. 221, 1989.

[13] W. Paton, *Proc. Roy. Soc. London*, vol. 154, p. 21, 1961.

[14] D. Sykes, M. Dowling and S. Charlton, *Mol. Pharmacol.*, vol. 76, p. 543, 2009.

[15] J. Nicholls, A. Martin, B. Wallace and P. Fuchs, From Neuron to Brain , Sunderland: Sinauer Assoc., 2001, p. 28 and further.

[16] H. Ing, *Science*, vol. 109, p. 264, 1949.

[17] A. Venkatakrishnan, X. Deupi, G. Lebon, C. Tate, G. Schertler and M. Babu, *Nature,* vol. 494, p. 185, 2013.

[18] B. Belleau and J. Puranen, *J. Med. Chem.*, vol. 6, p. 325, 1963.

[19] K. Haga, A. Kruse and H. Asada, *Nature*, vol. 482, p. 547, 2012.

[20] C. Czajkowski, C. Kaufmann and A. Karlin, *Proc. Natl. Acad. Sci. USA*, vol. 90, p. 6285, 1993.

[21] C. Cantor and P. Schimmel, Biophysical Chemistry, vol. 1, New York: W.H. Freeman and Co., 1980, pp. 245–247.

[22] P. Kukic, D. Farrell, L. McIntosh, B. García-Moreno, K. Jensen, Z. Toleikis, K. Teilum and J. Nielsen, *J. Am. Chem. Soc.*, vol. 135, p. 16968, 2013.

[23] A. Rubin, Fundamentals of Biophysics, USA: Wiley, 2014.

[24] R. Stephenson, *Brit. J. Pharmacol.*, vol. 11, p. 379, 1956.

[25] I. Komissarov, Elements of Receptor Theory in Molecular Pharmacology, Moscow: Medicine, 1969 (in Russian).

[26] P. Maguire, N. Tsai, J. Kamal, C. Cometta-Morini, C. Upton and G. Loew, *Eur. J. Pharmacol.*, vol. 213, no. 2, pp. 219–225, 1992.

[27] F. Dukhovich, M. Darkhovskii, E. Gorbatova and V. Kurochkin, Molecular Recognition: Pharmacological Aspects, New York: Nova Science Publishers, 2004.

[28] A. Funcke, R. Rekker, N. Ernsting, H. Tersteege and W. Nauta, *Arzneim. Forsch.*, vol. 9, p. 573, 1959.

[29] A. Burgen, *Br. J. Pharmacol.*, vol. 25, p. 4, 1965.

[30] F. Dukhovich, M. Darkhovskii, E. Gorbatova and V. Polyakov, *Chem. Biodiversity*, vol. 2, p. 354, 2005.

[31] F. Dreyer, K. Müller, K. Peper and R. Sterz, *Pflügers Arch.*, vol. 367, p. 115, 1976.

[32] D. Colquhoun and A. Hawkes, In: *Single-cannel recording*, E. Neher and B. Sakman, Eds., New York: Plenum Press, 1983, p. 135.

[33] A. Auerbach, *Neuropharmacology,* vol. 96, p. 150, 2015.

[34] A. Auerbach, *bioRxiv*, 2021, doi:10.1101/2021.12.18.473297.

[35] T. Nayak and A. Auerbach, *Proc. Natl. Acad.Sci. USA*, vol. 114, p. 11914, 2017.

[36] N. Fruentov, In: *Science results: Pharmacology and toxicology*, Moscow: VINITI, 1964, p. 41 (in Russian).

[37] A. Albert, Selective Toxicity, London: Chapman and Hall, 1983.

[38] H. Hartzell, *Nature*, vol. 291, p. 539, 1981.

[39] A. Brown and L. Birnbaumer, *Ann. Rev. Physiol.*, vol. 52, p. 197, 1990.

[40] E. Ariëns, *Arch. Int. Pharmacodynamie Therapie*, vol. 99, p. 32, 1954.

[41] C. Valant, K. Gregory, N. Hall, P. Scammels, M. Lew, P. Sexton and A. Christopoulos, *J. Biol. Chem.*, vol. 283, p. 29312, 2008.

[42] X. Chen, J. Klöckner, J. Holze, et al., *J. Med. Chem.*, vol. 58, p. 560, 2015.

[43] D. Volpato, M. Kauk, R. Messerer, M. Bermudez, G. Wolber, A. Bock, C. Hoffmann and U. Holzgrabe, *ACS Omega*, vol. 5, p. 31706, 2020.

[44] D. Volpe, G. McMahon Tobin, R. Mellon, G, et al., *Regulatory Tox. Pharmacol.*, vol. 59, p. 385, 2011.

[45] S. Head, *J. Phyisol. (Gr.Brit.)*, vol. 334, p. 441, 1983.

[46] A. Mayiski, N. Vedernikova, V. Chistyakov and V. Lakin, Biological Aspects of Addiction, Moscow: Medicine, 1982 (in Russian).

[47] C. Chavkin, J. McLaughlin and J. Celver, *Mol. Pharmacol.*, vol. 60, p. 20, 2001.

[48] J. Meyer, G. Del Vecchio, V. Seitz, N. Massaly and C. Stein, *Br. J. Pharmacol.*, vol. 176, p. 4510, 2019.

[49] J. Magnan, S. Paterson, A. Tavani and H. Kosterlitz, *Naunyn-Schmiedeberg's Arch. Pharmacol.*, vol. 319, p. 197, 1982.

[50] A. Bianchetti, D. Nisato, R. Sacilotto, M. Dragonetti, N. Picerno, A. Tarantino and L. Manara, *Life Sci.*, vol. 33 (suppl.1), p. 415, 1983.

[51] A. Misra, N. Vadlamani and R. Pontani, *J. Pharm. Pharmacol.*, vol. 30, p. 187, 1978.

[52] G. Henderson, *Br. J. Pharmacol.*, vol. 172, p. 260, 2015.

[53] C. Pert and S. Snyder, *Mol. Pharmacol.*, vol. 10, p. 868, 1974.

[54] C. Berrie, N. Birdsall, A. Burgen and E. Hulme, *Brit. J. Pharmacol.*, vol. 66, p. 470, 1979.

[55] G. Battaglia, M. Shannon and M. Titeler, *Life Sci.*, vol. 32, p. 2597, 1983.

11 Nonselective Drugs

11.1 AFFINITY PROFILES OF NEUROLEPTICS AND ANTIDEPRESSANTS

It is very important to improve drug selectivity to key receptors to eliminate undesirable side effects and incompatibility with other drugs. However, an important group of pharmaceuticals does not demonstrate high selectivity. This group includes neuroleptics and antidepressants, which are effective for the treatment of serious conditions such as psychosis and depression.

The affinity profiles of the following compounds to various neuroreceptors are presented in Table 11.1.

- Chlorpromazine

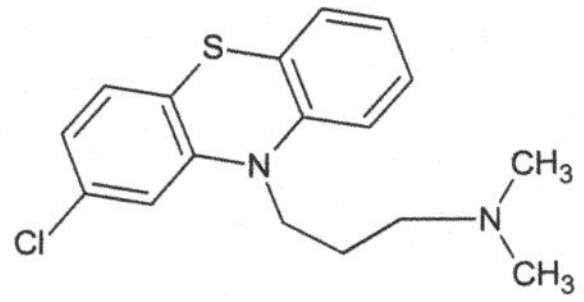

C-11.1 Chlorpromazine

- Flupentixol

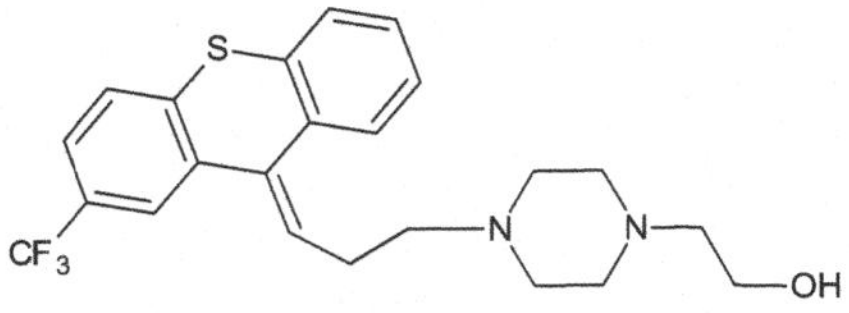

C-11.2 Flupentixol

- Fluphenazine

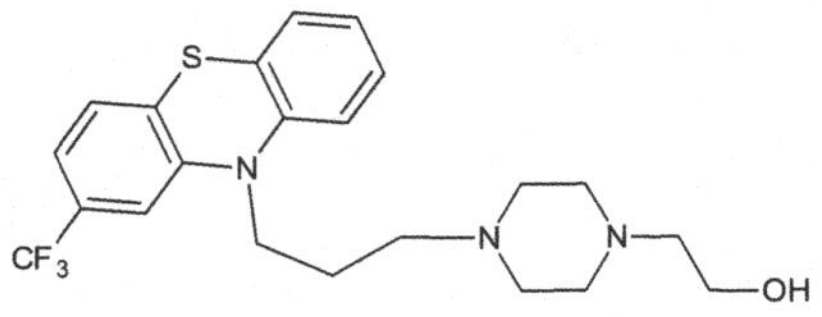

C-11.3 Fluphenazin

DOI: 10.1201/9781003366669-12

- Haloperidol

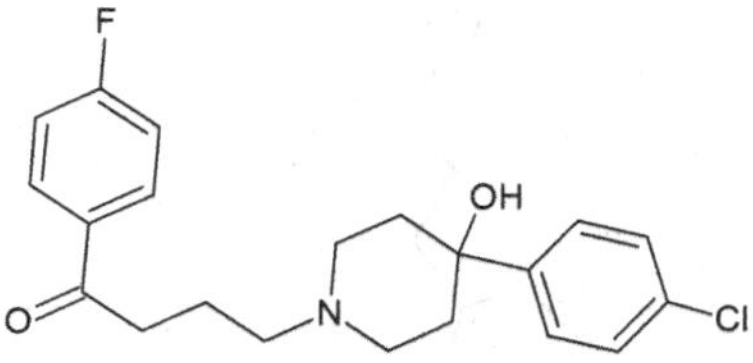

C-11.4 Haloperidol

- Pimozide

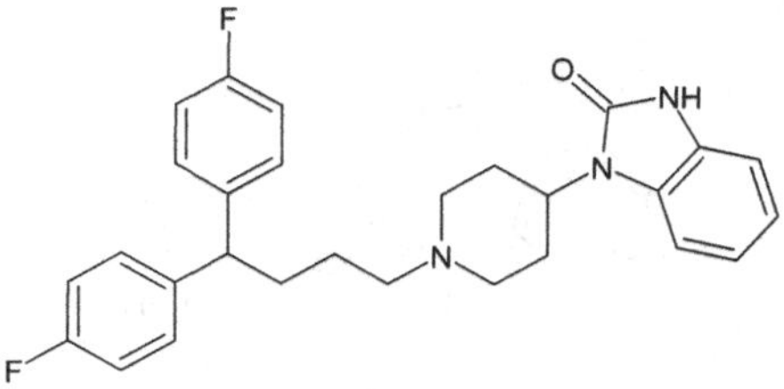

C-11.5 Pimozide

- Clozapine

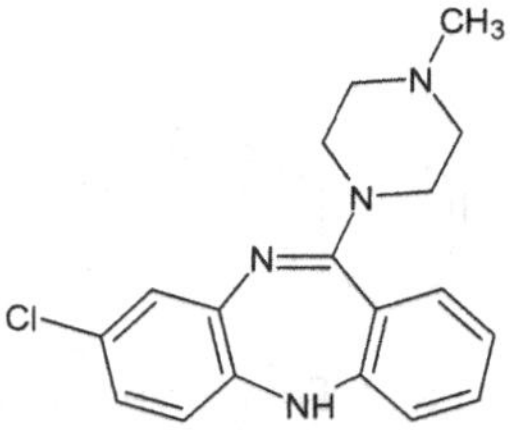

C-11.6 Clozapine

- Chlorprothixene

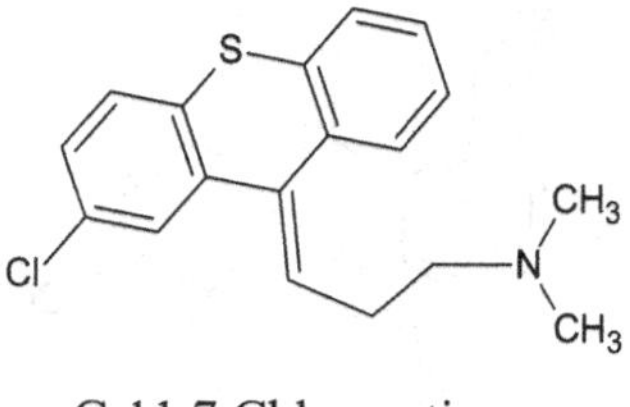

C-11.7 Chlorprotixene

- Perphenazine

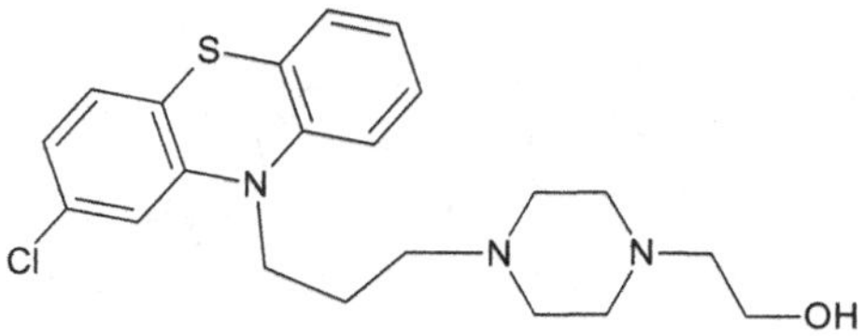

C-11.8 Perphenazine

- Mianserin

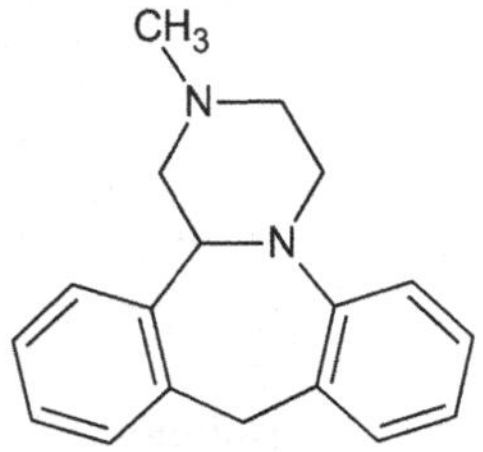

C-11.9 Mianserin

- Doxepin

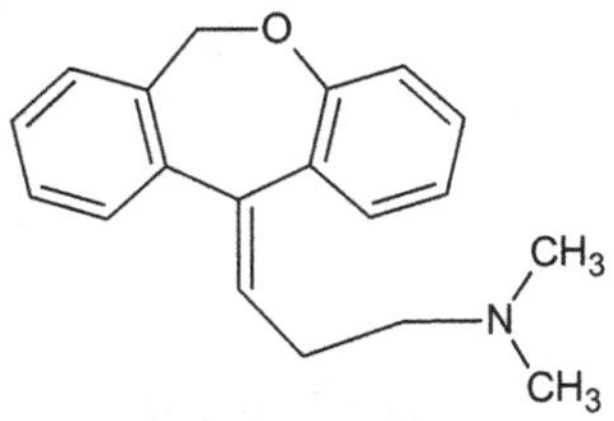

C-11.10 Doxepin

- Tranylcypromine

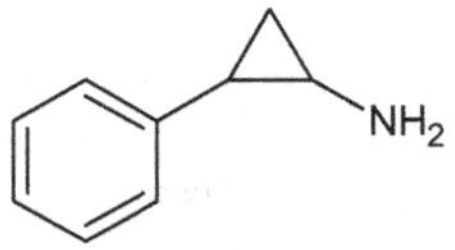

C-11.11 Tranylcypromine

- Sertraline

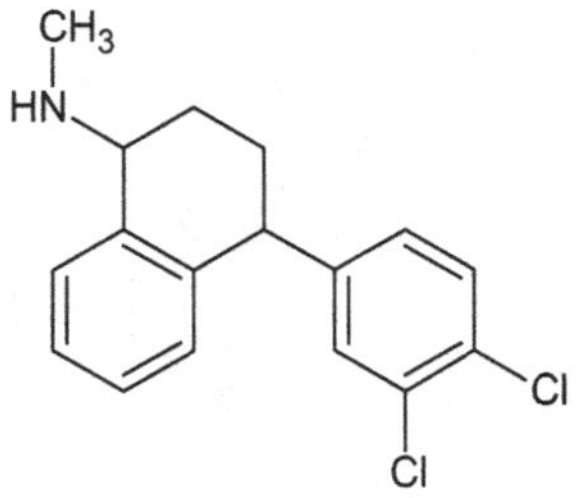

C-11.12 Sertraline

- Clomipramine

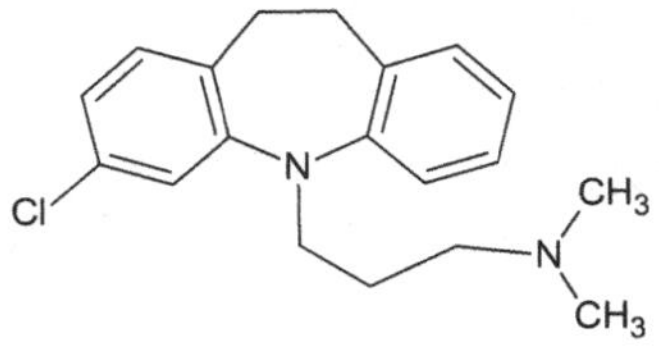

C-11.13 Clomipramine

TABLE 11.1
Receptor Affinity Profiles of Neuroleptics and Antidepressants

	Compound	mAChR	D_1	D_2	α_1	α_2	H_1	$5\text{-}HT_1$	$5\text{-}HT_2$
Neuroleptics	Chlorpromazine	$7.0\cdot10^{-8}$ [1]	$2.2\cdot10^{-8}$ [2]	$1.1\cdot10^{-9}$ [3]	$6.2\cdot10^{-9}$ [4]	$8.4\cdot10^{-7}$ [1]	$2.8\cdot10^{-8}$ [2]	$2.8\cdot10^{-6}$ [3]	$1.4\cdot10^{-9}$ [4]
	Flupentixol	$5.9\cdot10^{-8}$ [9]	$3.9\cdot10^{-8}$ [5]	$1.0\cdot10^{-10}$ [10]	$8.2\cdot10^{-9}$ [6]	$6.5\cdot10^{-7}$ [1]	$2.2\cdot10^{-8}$ [6]	$1.0\cdot10^{-6}$ [3]	$4.5\cdot10^{-9}$ [7]
	Fluphenazine	$6.8\cdot10^{-7}$ [13]	$5.0\cdot10^{-9}$ [5]	$5.0\cdot10^{-10}$ [14]	$1.3\cdot10^{-8}$ [6]	$5.8\cdot10^{-7}$ [15]	$1.5\cdot10^{-8}$ [6]	$1.2\cdot10^{-6}$ [3]	$2.5\cdot10^{-9}$ [8]
	Trifluoperazine	$7.4\cdot10^{-7}$ [13]	$2.0\cdot10^{-8}$ [9]	$1.2\cdot10^{-9}$ [3]	$6.7\cdot10^{-8}$ [6]	$1.4\cdot10^{-6}$ [17]	$1.4\cdot10^{-7}$ [2]	$2.3\cdot10^{-6}$ [4]	$4.0\cdot10^{-9}$ [8]
	Haloperidol	$2.7\cdot10^{-6}$ [10]	$5.0\cdot10^{-8}$ [5]	$1.0\cdot10^{-10}$ [10]	$1.4\cdot10^{-8}$ [6]	$1.8\cdot10^{-6}$ [5]	$2.6\cdot10^{-6}$ [2]	$2.8\cdot10^{-8}$ [4]	$9.5\cdot10^{-8}$ [7]
	Pimozide	$1.1\cdot10^{-6}$ [11]	$1.4\cdot10^{-6}$ [12]	$9.0\cdot10^{-10}$ [13]	$2.0\cdot10^{-8}$ [6]	$1.1\cdot10^{-6}$ [5]	$1.1\cdot10^{-6}$ [2]	$>1.0\cdot10^{-5}$ [14]	$8.0\cdot10^{-9}$ [8]
	Clozapine	$2.6\cdot10^{-8}$ [15]	$1.7\cdot10^{-7}$ [5]	$6.9\cdot10^{-9}$ [16]	$1.7\cdot10^{-8}$ [2]	$4.1\cdot10^{-6}$ [23]	$2.0\cdot10^{-8}$ [2]	$1.8\cdot10^{-6}$ [4]	$5.0\cdot10^{-9}$ [8]
	Chlorprothixene	$4.2\cdot10^{-8}$ [17]	$2.5\cdot10^{-7}$ [18]	$3.9\cdot10^{-9}$ [19]	-	$7.5\cdot10^{-7}$ [23]	-	$2.3\cdot10^{-7}$ [4]	$4.0\cdot10^{-10}$ [4]
	Perphenazine	$1.7\cdot10^{-6}$ [17]	$3.0\cdot10^{-7}$ [20]	$1.0\cdot10^{-9}$ [21]	$7.4\cdot10^{-9}$ [22]	$5.6\cdot10^{-7}$ [5]	$8.0\cdot10^{-9}$ [22]	$3.6\cdot10^{-6}$ [4]	$5.6\cdot10^{-9}$ [4]
Antidepressants	Imipramine	$9.0\cdot10^{-8}$ [17]	$2.0\cdot10^{-6}$ [5]	$2.0\cdot10^{-6}$ [22]	$5.1\cdot10^{-8}$ [23]	$8.7\cdot10^{-6}$ [1]	$1.5\cdot10^{-8}$ [24]	$1.4\cdot10^{-5}$ [24]	$2.6\cdot10^{-7}$ [25]
	Amitriptyline	$1.0\cdot10^{-8}$ [15]	$8.4\cdot10^{-8}$ [5]	$2.1\cdot10^{-7}$ [24]	$2.1\cdot10^{-8}$ [23]	$3.3\cdot10^{-7}$ [29]	$3.0\cdot10^{-9}$ [24]	$1.4\cdot10^{-6}$ [26]	$2.9\cdot10^{-8}$ [4]
	Mianserin	$8.0\cdot10^{-7}$ [17]	$2.6\cdot10^{-7}$ [5]	$3.0\cdot10^{-6}$ [27]	$5.3\cdot10^{-8}$ [28]	$4.2\cdot10^{-8}$ [5]	$3.0\cdot10^{-9}$ [24]	$1.0\cdot10^{-6}$ [26]	$7.0\cdot10^{-9}$ [4]
	Clomipramine	$3.5\cdot10^{-7}$ [24]	$1.6\cdot10^{-7}$ [5]	$1.9\cdot10^{-7}$ [22]	$2.7\cdot10^{-8}$ [24]	$3.0\cdot10^{-6}$ [24]	$3.3\cdot10^{-8}$ [24]	$1.2\cdot10^{-5}$ [24]	$2.7\cdot10^{-8}$ [4]
	Doxepin	$8.0\cdot10^{-8}$ [17]	$1.1\cdot10^{-7}$ [5]	$2.4\cdot10^{-6}$ [22]	$2.3\cdot10^{-8}$ [29]	$4.0\cdot10^{-6}$ [30]	$3.2\cdot10^{-10}$ [31]	$2.8\cdot10^{-7}$ [32]	$2.5\cdot10^{-8}$ [4]
	Tranylcypromine	-	$>1\cdot10^{-5}$ [5]	$>1\cdot10^{-5}$ [18]	$2.8\cdot10^{-6}$ [23]	$1.4\cdot10^{-7}$ [33]	-	$6.0\cdot10^{-6}$ [34]	$>1\cdot10^{-5}$ [34]
	Sertraline	$2.3\cdot10^{-7}$ [35]	-	$1.1\cdot10^{-5}$ [32]	$1.9\cdot10^{-7}$ [36]	$4.1\cdot10^{-6}$ [32]	$6.6\cdot10^{-6}$ [36]	$3.7\cdot10^{-6}$ [35]	$2.2\cdot10^{-6}$ [35]

K_d, M

11.2 NEUROLEPTICS

Currently, a rather well-justified dopamine hypothesis of psychosis (schizophrenia) is generally accepted [37-40]; according to the hypothesis, psychosis is caused by hyperactivity of dopamine neurons located in the central nervous system (CNS). Psychosis symptoms include psychomotor agitation and its acute forms, namely violence, unmotivated aggression, hallucinations, suicide attempts, and other manifestations. Neuroleptics block dopamine receptors, thereby reducing their reactivity. D_2 dopamine receptors are the main targets of neuroleptics. A correlation exists between the antipsychotic activity of neuroleptics and their affinity to D_2 receptors [41, 42]. Dopamine receptors are classified depending on the differences in the activation of adenylate cyclase by dopamine—the D_1 receptors activate this enzyme when compared with the D_2 receptors [43]. Other mediators in addition to dopamine as well as exogenous substances also play a significant role in psychosis development. For example, haloperidol demonstrates a relatively lower affinity to mAChR than other neuroleptics. The emerging symptoms of extrapyramidal disorders can be alleviated by prescribing haloperidol with m-anticholinergic agents. Neuroleptics display very low affinity to the receptors of the inhibitory mediator of gamma-aminobutyric acid (GABA). For example, clozapine C-11.6 and haloperidol C-11.4 have affinity for GABA receptors of $K_d > 10^{-4}$–10^{-5} M [44]. It is likely that GABA can have an antipsychotic effect. The effect of neuroleptics on CNS functions is not associated with the blockade of H_1-histamine receptors [45-a]. Accumulating information is available on the effect of neuroleptics on adrenergic, serotoninergic, GABA-ergic, cholinergic, and other neuromediator processes [46-a]. Mediators affect the neuron membrane potential through postsynaptic receptors. The membrane adds up excitatory signals and subtracts inhibitory signals reaching the membrane through numerous synaptic contacts. The neuron function involves selection of the strongest (most informative) signal of the action potential, which is transmitted through axon according to the all-or-nothing principle. The local graduated potentials are the decaying ones and are propagated only within the limits of certain regions of the neuron membrane. Action potential (spike) is propagated without decay and reaches the target cell either through another neuron or through a muscle cell. The rate of action potential propagation is approximately 100 m/s [47].

11.3 DOPAMINE AND NEURON ACTION POTENTIAL

Schematic representation of a neuron excitation under normal (A) and pathological (B) conditions is shown in Figure 11.1.

V represents the membrane potential, V_R represents the resting membrane potential, V_R^* represents the resting membrane potential under pathological conditions, V_E represents the threshold of excitation, and t represents time.

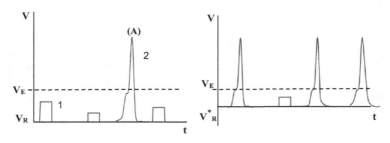

FIGURE 11.1 Graduated potentials (1) and action potential (2) of an average neuron under normal (left) and pathological (right) conditions.

The average resting potential of a neuron is−70 mV (the inner membrane surface carries a negative charge). Action potential emerges when depolarization reaches an average level of −40 mV, which is termed "excitation threshold" [47].

The increase in dopamine concentration in the synapse decreases the resting potential (in absolute value). Hence, the situation arises when the stimuli that did not lead to a spike under normal conditions now cause it. This type of signal distortion is the underlying cause of psychosis characterized by excitation of the nervous system. Under conditions of subthreshold depolarization (V_R^*), action potentials can be caused not only by dopamine but also by other excitatory mediators.

We suppose that the increase in noradrenaline, serotonin, and acetylcholine activity occurs not because of the increase in their concentration in the synaptic gap but because of the decrease in the resting potential caused by dopamine. On the contrary, neuroleptics help decrease their activity by blocking noradrenaline, serotonin, and acetylcholine receptors.

11.4 SIGNIFICANT AND INSIGNIFICANT RECEPTORS

Considering nonselective drugs, it is useful to determine which receptors can be blocked to achieve significant therapeutic effects with respect to the key receptor and which receptors need not be blocked.

Drug selectivity δ can be expressed through partial blockade of two receptors in the following ways:

$$\delta = i_1 - i_2$$

where i_1 represents the degree of blockade of the receptor mediating the main effect of the compound, and i_2 indicates the degree of blockade of the second receptor associated with the side effect.

In the case of absolute selectivity, $\delta = 1$. In the case of partial selectivity, when both receptors are bound at a certain degree, $\delta < 1$. The δ value depends on concentration of the compound [I]. Partial blockade can be expressed using the following equations:

$$i_1 = \frac{[I]}{Kd_1 + [I]}; \quad i_2 = \frac{[I]}{Kd_2 + [I]}.$$

Hence, under conditions of constant [I]:

$$\frac{Kd_2}{Kd_1} = \frac{i_1(1-i_2)}{i_2(1-i_1)}.$$

When the values of Kd_1 and Kd_2 differ by tenfold, the degree of blockade of the first-type receptors ($i_1 = 0.5$) would be 50%, and the degree of blockade of the second-type receptors would be 9%. The 75% blockade of the first-type receptors ($i_1 = 0.75$) would result in 23% blockade of the second ones. When the difference in K_d is 100-fold or more, the blockade of the second-type receptors becomes insignificant: 3% in the case of 75% blockade of the first-type receptors and 1% in the case of 50% blockade of the first-type receptors. If one considers the blockade of less than 5% of receptors as negligible, then the blockade of α_1-adrenoreceptors, mAChR receptors, H_1 histamine receptors, and 5-HT$_2$-receptors of 5-hydroxytryptamine (serotonin) by neuroleptics is significant with respect to the D$_2$ receptors. Blockade of the remaining receptors shown in Table 11.1 can be disregarded.

If the drug demonstrates strong affinity to two or several receptors, then there is nothing surprising in the fact that it alleviates symptoms associated with the effect on these receptors and causes undesirable side effects. For example, the antihistamine drug cyproheptadine demonstrates high activity toward 5-HT_2-, mAChR, and D_2-receptors in addition to H_1-receptors (see Table 7.4 in Chapter 7). In addition to its use in the treatment of allergic reactions (antihistamine action), cyproheptadine is effective in treating migraines (as a 5-HT_2-receptor antagonist), blocking hypersecretion of somatotropin in acromegaly, and stimulating the secretion of adrenocorticotropic hormone in Itsenko-Cushing syndrome (caused by the effect on D_2 receptors). At the same time, this preparation can cause dry mouth conditions, drowsiness, and sleepiness (anticholinergic action), and therefore, it is not recommended for administration to patients during working hours if their work requires being on high alert and strenuous physical activity [48].

11.5 NEUROLEPTICS AND ANTIDEPRESSANTS: STRUCTURAL DIFFERENCES DETERMINING THE TYPE OF PHARMACOLOGICAL ACTIVITY

The range of affinity of neuroleptics and antidepressants toward monoamine receptors is presented in Table 11.1. These receptors have attracted considerable interest recently in association with the mechanism of action of these medications predominantly toward the dopamine, noradrenaline, serotonin, and acetylcholine receptors. The neuroleptics are represented by several major structure types: tricyclic derivatives of phenothiazine (chlorpromazine, fluphenazine, and trifluoperazine), derivatives of butyrophenone (haloperidol), thioxanthene (chlorprothixene), dibenzodiazepine (clozapine), and others.

Minor changes in the structures of the tricyclic neuroleptics involving replacement of the central six-membered ring by the seven-membered one cause transition in the therapeutic from the category of neuroleptics to the category of antidepressants. According to Albert's observations, these two activities never occur together in any of the tricyclic compounds. Based on his suggestion, the type of pharmacological activity of these compounds is determined primarily by the bending angle α of the cyclic system. The relatively flat molecules ($\alpha \sim 25°$) mostly demonstrate the properties of neuroleptics, such as chlorpromazine and trifluoperazine (Figure 7.6), while those compounds with more bent skeletons ($\alpha \sim 55-65°$) demonstrate the properties of antidepressants (imipramine (Figure 5.2) and clomipramine) [49]. The increase in the α angle leads to the decrease in the compound's affinity to the D_2 receptor by 1,000-fold to 10,000-fold compared with neuroleptics while not affecting the affinity to other receptors.

The antidepressants pirlindole (C-**11.14**), tetrindole (C-**11.15**), and moclobemide (C-**11.16**) are reversible inhibitors of monoamine oxidase (MAO). The tricyclic compounds amitriptyline and imipramine (Figure 5.2) are inhibitors of neuronal reuptake of neurotransmitters [46-b].

Pirlindole

C-11.14 Pirlindole

Tetrindole

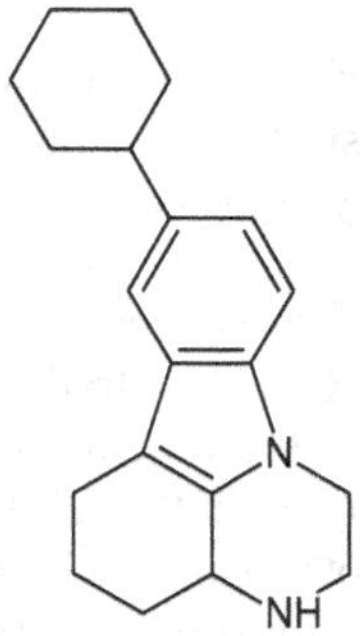

C-11.15 Tetrindole

Moclobemide

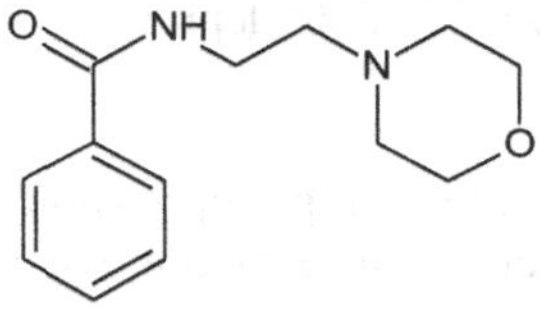

C-11.16 Moclobemide

The MAO (type A) inhibition constants K_i for moclobemide, tetrindole, and pirlindole are 110 nM [50], 400 nM [51], and 230 nM [52], respectively, with these compounds having a low affinity to the D_2 dopamine receptor. Affinity of moclobemide to this receptor is characterized by the K_d value of 18200 nM [50].

11.6 LITHIUM

Lithium (lithium carbonate and lithium oxybutyrate) is used for the prophylaxis and treatment of maniac and manic-depressive psychoses [45]. The mechanism of lithium action has not been elucidated. It is likely that the action of lithium is associated with the properties of the Li^+ hydration shell. This suggestion arose from the fact that sodium, potassium, beryllium, and aluminum are not expressing lithium-like action. Ionic radii of these elements are as follows: Li^+, 0.60 Å; Mg^{2+}, 0.65 Å; Be^{2+}, 0.31 Å; Al^{3+}, 0.50 Å; Na^+, 0.95 Å; K^+, 1.33 Å [53]; hence, the therapeutic properties of lithium are not determined by its ionic radius. Hydration shells of these ions are significantly different from the Li^+ hydration shells. In particular, the first layer of the Li^+ hydration shell contains four water molecules [54], whereas the first layer of Na^+ [54] and Mg^{2+} [55] hydration shells includes six water molecules.

While consideration of the Li^+ hydration shell can be limited to the first layer of water molecules, the next hydration layer should be considered for the Mg^{2+}, Be^{2+}, and Al^{3+} ions, as these ions carry higher charges than those of the lithium ion. Because of the higher ionic radii of Na^+ and K^+ ions, their first hydration shells have larger radii than those of Li^+.

It is likely that the compactness of the hydrated lithium ion makes it into a specific ligand for a certain receptor or enzyme (probably MAO ligand) that has an anionic center in the active site with the size compatible with that of the hydrated lithium ion.

11.7   ANTIDEPRESSANTS: MODERN VIEWS ON MOLECULAR MECHANISMS OF ACTION

According to the hypothesis on the role of biogenic amines in psychopathology, depression occurs because of the deficiency of neurotransmitters in those sections of the brain that are responsible for sleep, excitability, appetite, psychomotor activity, and adequate response to incoming information.

The main cause of depression is the low concentration of noradrenaline and serotonin in the synaptic cleft. The decrease in the mediator concentration is caused by the disruption in their synthesis. It is a generally accepted view that the function of an antidepressant involves the increase in noradrenaline and serotonin concentrations in the CNS synapses through the inhibition of MAO enzyme that inactivates monoamine neurotransmitters by oxidative deamination, as well as through the blockade of neuronal reuptake of these transmitters by presynaptic terminals [45-a, 46-b, 47, 49]. The effect of antidepressants is evident 5–10 days after dose initiation, on average.

11.8   ANTIDEPRESSANTS: HYPOTHESIS AND CONCERNS

The existing hypothesis on the mechanism of antidepressant action is not well justified based on the following considerations:

1. According to this hypothesis, alleviating depression with noradrenaline and serotonin should be mediated by the effect of these mediators on postsynaptic receptors. However, most antidepressants block their signals, as they demonstrate high affinity (K_d ~10^{-8} M) to α_1- and 5-HT$_2$-receptors through which the noradrenaline and serotonin signals are transmitted to the CNS. It seems that antidepressants should not interfere with the effects of these mediators.
2. The antidepressants listed in Table 11.1 mediate the increase in noradrenaline and serotonin concentrations in the synapses [56]. These include MAO inhibitor (tranylcypromine) and blockers of noradrenaline and serotonin reuptake that prevent their diffusion from the synaptic cleft (imipramine, sertraline, amitriptyline, and others). However, the group of antidepressants also includes compounds such as mianserin that do not inhibit MAO and do not block the mediator reuptake [46-b].
3. Many of the antidepressants demonstrate similar (and high) affinity to α_1- and 5-HT$_2$-receptors as neuroleptics (Table 11.1). This is rather confusing as the functions of these compounds are opposite: neuroleptics decrease the activity of neurotransmitters in the CNS, whereas antidepressants do not interfere with their activity.

11.9   ALTERNATIVE VIEWS ON THE MECHANISM OF ANTIDEPRESSANT ACTION

The inconsistencies described in the previous sections are mainly because the main role of antidepressant action is assigned to noradrenaline and serotonin. However, as demonstrated above, the action of monoamine neurotransmitters on the specific receptor is actively blocked by antidepressants. Here, it is worth mentioning the following feature of antidepressants independent on their numerous scaffolds: without any exception, they demonstrate much lower affinity to D_2 dopamine receptors than neuroleptics. This implies that antidepressants do not interfere with the effects of dopamine. Here again, dopamine, which, according to the modern views on depression development and mechanisms of antidepressant action, has not gained extensive attention, comes under the spotlight. We suggest that dopamine plays a crucial role in either depression or psychosis: deficiency of dopamine in depression and excess in psychosis.

One can argue that nothing changes with the replacement of noradrenaline and serotonin by dopamine, considering that there are no direct verifications of the hypothesis. However, a few lines of evidence should be mentioned that follow from the hypothesis on the leading role of dopamine in the development of depression:

1) the efficacy of substances that increase the dopamine concentration in the CNS would be worth investigating
2) the requirement of low affinity of the compound to the D_2 receptor must be considered during the search for new antidepressants
3) considering that the MAO type A deactivates dopamine, continuing the search for the high inhibiting ability of these compounds toward this enzyme would be beneficial
4) it is likely that the decrease in the affinity of the compounds to α_1- and 5-HT_2-receptors could increase the synaptic activity of noradrenaline and serotonin, the effect of which, in our opinion, is concurrent to the action of dopamine.

11.10   CONCLUSION

While the mechanisms of neuroleptic action are justified and generally accepted in the literature, the same cannot be confirmed for the pharmacological effects of antidepressants. The fact that all antidepressants display low affinity to the D_2 dopamine receptor is in favor of the key role of dopamine in depression development and depression treatment strategies, considering the much higher affinity of antidepressants to the receptors of noradrenaline (α_1-receptors) and serotonin (5-HT_2 receptors).

As mentioned above, the action of currently available antidepressants is associated with either MAO inhibition or neuronal reuptake blockade, although antidepressants that do not demonstrate these properties are also effective. Existing alternative hypotheses suggest the increase in the number of receptors and increase in their sensitivity under the action of antidepressants [45-a].

Aside from the complex mechanisms of psychopathology, the large number of suggested hypotheses is attributed to the lack of adequate animal models of depression as well as an insufficient amount of experimental data on the participation of various receptors and enzymes in the genesis of the depressive states. Meanwhile, it is expected that the problem will be solved.

Paul Erlich wrote: "We should strive to solve problems at the molecular level" (in German original: "... Wir mussen uns bemuhen, naher in das Wesen des Prozesses einzudringen, und de sedibus et causis pharmacophorum Einsicht zu gewinnen" [61]). We have followed this advice in this chapter while investigating the interactions of psychotropic medications with their receptors.

REFERENCES

[1] M. Terai, S. Usuda, I. Kuroiwa, O. Noshiro and H. Maeno, *Japan J. Pharmacol.*, vol. 33, p. 749, 1983.

[2] S. Peroutka and S. Snyder, *Am. J. Psych.*, vol. 137, p. 1518, 1980.

[3] S. Mason and G. Reynolds, *Eur. J. Pharmacol.*, vol. 221, p. 397, 1992.

[4] T. Wander, A. Nelson, H. Okazaki and E. Richelson, *Eur. J. Pharmacol.*, vol. 132, p. 115, 1986.

[5] G. Faedda, N. Kula and R. Baldessarini, *Biochem. Pharmacol.*, vol. 38, p. 473, 1989.

[6] S. Hill and M. Young, *Eur. J. Pharmacol.*, vol. 52, p. 397, 1978.

[7] M. Quik, L. Jverssen and A.M.A. Larder, *Nature,* vol. 274, p. 513, 1978.

[8] H. Meltzer and J. Hash, *Pharmacol. Rev.*, vol. 43, p. 587, 1991.

[9] H. Meltzer, S. Matsubara and J.-C. Lee, *J. Pharmacol. Exp. Ther.*, vol. 251, p. 238, 1989.

[10] S. Snyder, G. Grenberg and H. Yamamura, *Arch. Gen. Psychiat.*, vol. 31, p. 58, 1974.

[11] P. Laduron and J. Leysen, *J. Pharm. Pharmacol.*, vol. 30, p. 120, 1978.

[12] P. Andersen, F. Gronvald and J. Jensen, *Life Sci.*, vol. 37, p. 1971, 1985.

[13] P. Jenner and C. Marsden, *Acta Psychiatr. Scand. Suppl.*, vol. 311, p. 109, 1984.

[14] B. Roth, R. Ciaranello and H. Meltzer, *J. Pharmacol. Exp. Ther.,* vol. 260, p. 1361, 1992.

[15] S. Snyder and H. Yamamura, *Arch. Gen. Psychiat.,* vol. 34, p. 236, 1977.

[16] N. Eltayer, G. Kilpatrick, H. Van de Waterbeemd, B. Testa, P. Jenner and C. Marsden, *Eur. J. Med. Chem.,* vol. 23, p. 173, 1988.

[17] S.-C. Lin, K. Olson, H. Okazaki and E. Richelson, *J. Neurochem.,* vol. 46, p. 274, 1986.

[18] D. Burt, I. Creese and S. Snyder, *Mol. Pharmacol.,* vol. 12, p. 800, 1976.

[19] J. Leysen, W. Gommeren and P. Laduron, *Biochem. Pharmacol.,* vol. 27, p. 307, 1978.

[20] W. Billard, V. Ruperto, G. Crosby, L. Iorio and A. Barnett, *Life Sci.,* vol. 35, p. 1885, 1984.

[21] T. Lee and P. Seeman, *Soc. Neurosci. Abstr.,* vol. 5, p. 653, 1979.

[22] E. Richelson and A. Nelson, *Eur. J. Pharmacol.,* vol. 103, p. 197, 1984.

[23] D. U'Prichard, D. Greenberg and S. Snyder, *Mol. Pharmacol.,* vol. 13, p. 454, 1977.

[24] H. Hall and S.-O. Ogren, *Eur. J. Pharmacol.,* vol. 70, p. 393, 1981.

[25] J. Leysen, C. Niemegeers, J. Van Nueten and P. Laduron, *Mol. Pharmacol.,* vol. 21, p. 301, 1982.

[26] S. Peroutka, *J. Neurochem.,* vol. 47, p. 529, 1986.

[27] M.-P. Martres, N. Sales and M.-L. Bouthenet, *Eur. J. Pharmacol.,* vol. 118, p. 211, 1985.

[28] D. Hoyer, P. Vos, A. Closse, A. Pazos, J. Palacios and H. Davies, *Naunyn-Schmeideberg's Arch. Pharmacol.,* vol. 335, p. 226, 1987.

[29] D. U'Prichard, D. Greenberg, P. Sheehan and S. Snyder, *Science,* vol. 199, p. 197, 1978.

[30] H. Rommelspacher, S. Strauss, E. Fahndrich and H. Haug, *J. Neural. Transm.,* vol. 69, p. 85, 1987.

[31] L. Toll and S. Snyder, *J. Biol. Chem.,* vol. 257, p. 13593, 1982.

[32] B. Cusak and E. Richelson, *J. Recept. Res.,* vol. 13, p. 123, 1993.

[33] P. Timmermans, J. van Kemenade, Y. Harms, G. Prop, E. Graf and P. Van Zwieten, *Arch. Int. Pharmacodyn. Ther.,* vol. 261, p. 23, 1983.

[34] L. Toll, I. Berzetei-Gurske, W. Polgar, S. Brandt, I. Adapa, L. Rodriguez, R. Schwartz and D. Haggart, *NIDA Res. Monogr.,* vol. 178, p. 440, 1998.

[35] M. Owens, W. Morgan, S. Plott and C. Nemeroff, *J. Pharmacol. Exp. Ther.,* vol. 283, p. 1305, 1997.

[36] M. Owens, D. Knight and C. Nemeroff, *Encephale (in French),* vol. 28, p. 350, 2002.

[37] A. Horn and S. Snyder, *Proc. Natl. Acad. Sci. USA,* vol. 68, p. 2325, 1971.

[38] A. Carlsson and M. Lundqvist, *Acta Pharmacol. Toxicol.,* vol. 20, p. 140, 1963.

[39] J. Haracz, *Schizophrenia Bull.,* vol. 8, p. 438, 1982.

[40] E. Meisenzahl, G. Schmitt and H. Moller, *International Review of Psychiatry,* vol. 19, no. 4, p. 337, 2007.

[41] J. Creese, D. Sibley, M. Hamblin and S. Left, *Annu. Rev. Neurosci.,* vol. 6, p. 43, 1983.

[42] F. Huho, Neurochemistry (Russian transl.), Moscow: Mir, 1990, p. 249.

[43] J. Kebabian and D. Calne, *Nature,* vol. 277, p. 93, 1979.

[44] H. Schoemaker, Y. Claustre, D. Fage, L. Rouquier, K. Chergui, O. Curet, A. Oblin, F. Gonon, C. Carter, J. Benavides and B. Scatton, *J. Pharm. Exp. Therap.,* vol. 280, p. 83, 1997.

[45] D. Laurence and P. Bennett, Clinical Pharmacology, Edinburgh: Churchill Livingstone, 1987, pp. (a) 91, (b) 114.

[46] M. Mashkovskii, Lekarstvennye Sredstva (Drugs), vol. 1, Moscow: Novaya Volna, 2000, pp. (a) 53; (b) 106 (in Russian).

[47] J. Nicholls, A. Martin, B. Wallace and P. Fuchs, From Neuron to Brain, Sunderland: Sinauer Assoc., 2001, p. 28 and further.

[48] Drug Encyclopedia, RLS, 2006, p. 932 (in Russian).

[49] A. Albert, Selective Toxicity, London: Chapman and Hall, 1983.

[50] V. Badavath, I. Baysal, G. Ucar, B. Sinha and V. Jayaprakash, *ACS Med. Chem. Lett.,* vol. 7, p. 56, 2015.

[51] A. Medvedev, A. Kirkel, N. Kamyshanskaya, et al., *Biochem. Pharm.,* vol. 47, p. 303, 1994.

[52] E. Schraven and R. Reibert, *Arzneimittelforschung.,* vol. 34, p. 1258, 1984.

[53] L. Pauling, Nature of the Chemical Bond, 3rd ed., New York: Cornell University Press, 1960.

[54] J. Mahler and I. Persson, *Inorg. Chem.,* vol. 51, p. 425, 2012.

[55] F. Bruni, S. Imberti, R. Mancinelli and M. Ricci, *J. Chem. Phys.,* vol. 136, p. 064520, 2012.

[56] A. Ariens and A. Simonis, In: *Beta-adrenoceptor Blocking Agents,* P. Saxena and R. Forsyth, Eds., Amsterdam: Elsevier/North-Holland, 1976.

[57] A. Hancock and C. Marsh, *Mol. Pharmacol.,* vol. 26, p. 439, 1984.

[58] S. Enna, J. Bennett, D. Burt, I. Creese and S. Snyder, *Nature*, vol. 263, p. 338, 1976.

[59] B. Syoholm, R. Voutilainen, K. Luomala, J.-M. Savola and M. Scheinin, *Eur. J. Pharmacol.*, vol. 215, p. 109, 1992.

[60] R. Hornung, P. Presek and H. Glossmann, *Naunyn-Schmeideberg's Arch. Pharmacol.*, vol. 308, p. 223, 1979.

[61] F. Himmelweit, Ed., The Collected Papers of Paul Ehrlich, vol. 3 (Chemotherapy), London: Pergamon Press, 1960, p. 169.

12 Search Paths for Novel Therapeutic Drugs

12.1 SEARCH STRATEGIES FOR NEW DRUGS: OBJECTIVES AND MEANS

In recent years, the intensity of the development of new drugs has increased considerably. At the same time, the costs involved in the drug design industry have also increased dramatically. The plot shown in Figure 12.1 depicts the substantial increase in the average costs of developing a new drug (for each drug entering clinical studies) from the 1960s to the end of the 1990s for the largest pharmaceutical companies.

Estimates of the median costs of R&D of new drugs released on the market in the 2000s and 2010s have increased significantly, amounting to a range of US$985 million and US$1395 million per successful drug [2, 3].

Starting from the full cost for developing a new drug to bringing it to the market, less than 26% is spent on the drug design pipelines (R&D) [4], which is still a huge amount considering the total cost.

Clearly, all these data suggest that the process of developing a new pharmaceutical drug and bringing it to the market is long and expensive. Currently, approximately one or two decades may be required for potent drug preparation from the first synthesis in the laboratory to the start of its clinical use (on average, from 9 to 12 years) [5]. Investigation of the pharmacokinetics, metabolism, and toxicity of the new drug takes a large amount of time during this period. Screening for potential drug candidates based on potent activity takes, on average, from 3 to 5 years.

Special requirements are imposed on the toxicity of drugs. Strict government-regulated toxicity testing of drug candidates was introduced because of the thalidomide tragedy. During 1960–1961, the outbreak of phocomelia (malformation of the limbs) occurred in West Germany. The Ministry of Health of the Federal Republic of Germany reported that approximately 10 000 infants were born with this disease, from which only half survived. Many of those who survived continued to live with malformed limbs, eyes, heart, and gastrointestinal and urinary tracts. The pathology was caused by exposure of pregnant women to thalidomide, which was marketed widely as an over-the-counter new sedative and sleep medication. Phocomelia also occurred in other countries including Australia but at a smaller scale. Thalidomide is not the only medication that caused severe pathology. More than 15 preparations have been cited in the medical literature in the 1980s, the administration of which was associated with high mortality, but, to a lesser degree, than that in the case of thalidomide [6]. The significantly higher number of preparations caused negative side effects. A paradoxical situation had emerged when the drug was approved for clinical use based on experiments conducted with laboratory animals, while the toxicity for humans could be determined only after several years of use in the clinical setting. Only one solution to resolve this situation exists: the mechanism of action of the medication must be elucidated.

A schematic representation of the drug development pipeline is shown in Table 12.1.

DOI: 10.1201/9781003366669-13

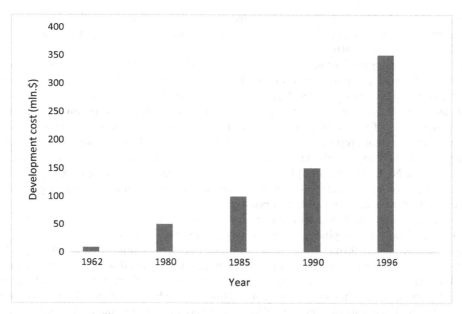

FIGURE 12.1 Cost increase of a new drug development during 1960–1996 (Source: [1]).

TABLE 12.1
Steps in the Development of a New Drug

Step 1
Target selection. Selection of potential drug candidates: natural compounds, new drug-like molecules, and virtual
  compounds
Step 2
Synthesis and physicochemical examination of the compounds
Step 3
Investigation (prediction) of biological activity
Step 4
Toxicity screening of the compounds (Phase 1 of preclinical studies)
Step 5
Investigation of the specific biological activity (Phase 2 of preclinical studies)
Step 6
Clinical studies

Step 1: Step 3 is discussed with far more details in [7].
The available information suggests that the number of compounds decreases by one order of magnitude on transition from one step to the next (Table 12.1), and the cost of the steps increases accordingly. Studies conducted in the USA demonstrated that bringing to market 68 drugs arbitrarily selected by the pharmaceutical company involved spending of US$802 million on the investigation of preparations rejected in Steps 2–4 [8].

Recent progress in the fields of biochemistry, synthetic chemistry, and combinatorial chemistry has made pharmaceutical companies to pay more attention to the development of prescreening techniques for testing and, thus, bringing down costs of drug development [9].

Financing of drug development has increased many times from the 1990s. High-throughput screening (computer-simulated molecular targeting based on the biochemical pathway), new chemical entities (NCE; referring to a new drug molecule or compound that has not yet been tested on humans), and 3DMD (3D molecular docking) are among the most advanced screening approaches for new drugs [10].

Despite the growing body of research, currently, there is a need for a methodology in the search for new drugs because computer modeling has not been able to justify the investments, meet bright expectations, and force pharmaceutical companies to pursue alternative investment strategies [11].

NCE and 3DMD, as well as combined methods, are considered most promising [12, 13]. The commercial success of this strategy is related to the fact that pharmaceutical companies use clinical data for the original preparation while producing generics but receive approval and patent for the generic as a "new use for the known compound." Hence, development of a screening system (according to the NCE classification) based on a model that considers, to the maximum extent possible, the features of human metabolism using biosensor technology is very interesting scientifically and a very important task from a practical point of view.

The screening methods based on the chemical combination of small fragments, which have a relatively low affinity to the pharmacological target separately, wherein the compounds from the fragments are produced through chemical combination, demonstrate a relatively high affinity, making it a lead compound for further optimization, and have gained wide acceptance during the 1990s-2000s (fragment-based drug design) [14, 15]. These methods represent one of the variants of the NCE approach. Another variant is "scaffold hopping," wherein structurally novel compounds are sought by modifications of somewhat determined core chemical structure in a series of known active compounds [16, 17, 18]. Core chemical structure can be loosely but reliably defined in many series of compounds. Another set of methods is based on rapid screening of molecular libraries of the compounds covalently bound to DNA fragments (DNA-encoded libraries) [19, 20, 21]. In this case, DNA is used as a label (tag) for rapid identification of molecules binding to the protein target.

There are two levels of the drug design complexity: development of a novel type of preparation and improvement of the existing ones. Novel drug types imply the development of compounds that offer cure for those diseases that are considered incurable at present. As expected, any approach allowing rapid identification of novel drugs allows a decrease in the rate of growing costs. Theoretical drug design methods must significantly narrow down the range of compounds that could be recognized as potential drugs.

The main objectives that should be followed during the development of new pharmaceuticals are summarized in Table 12.2.

12.2   SEARCH FOR DRUGS WITH NEW-TYPE ACTION

Discoveries of the "miracle drugs"—antibiotics, neuroleptics, sleeping pills, analgesics, vitamins, immunostimulators, some anticancer preparations, and many others—are considered as the greatest achievements of scientists. Unlike chemotherapeutic agents, the majority of the pharmacodynamics drugs do not eliminate the cause of the disease but rather alleviate the symptoms of the disease, thus helping the host to recruit all inner reserves to fight it. Nevertheless, expectations are high globally that remedies to cure diseases such as cancer, AIDS, Parkinson's disease, and Alzheimer's disease will be discovered. It is evident from historical lessons that the discoveries of the fundamentally new drugs that paved the road for a new class of drugs occurred serendipitously, although these discoveries were made as a result of observations that required certain scientific knowledge and intuition [22]. In many cases, natural compounds served as a source for drugs. Although this source cannot be, by any means, considered exhausted, it cannot be considered bottomless either. At present, any strategy for predicting the chemical structures of novel drugs does not remain unnoticed. Automated search systems have attracted specific attention.

The higher the diversity of the structures in the considered set of compounds with a prescribed number of them in the set, the higher is the probability to find a valuable compound. As a rule, a

TABLE 12.2
Objectives and Methods of Drug Development

Objectives	Methods
1. Development of novel types of drugs	Random search. De novo drug design.
2. Development of drugs with improved properties	Modification of the base structure. Combination of novel small fragments in fragment-based drug design.
2.1. Increase in drug activity (dose reduction)	Determination of optimal topography of the molecule with the help of conformational analysis.
	Introduction of anchor groups into the molecule that do not interfere with the recognition signature.
2.2. Increase in the selectivity. Reduction in side effects	Removal of nontarget receptor pharmacophores from the molecule and other changes of the compound structure directed to the increase in the affinity difference between target and nontarget receptors.
2.3. Development of drugs with prolonged action [1]	Increase in the lifetime of the complex of the compound with the receptor through increasing affinity. Improving resistance of the compound to biodegradation. Development of prodrugs.
2.4. Reducing tolerance	Increasing the number of drug variations. Development of preparations affecting biochemical processes forming tolerance.
2.5. Personalized drugs	Achieving pharmacokinetic compatibility with other medications administered to the patient. Modification of the compound structure to produce more efficient drugs for the patient treated for other conditions that consider physiological peculiarities of the individual.
2.6. Optimization of pharmacokinetic properties. Targeted drug delivery.	Varying physicochemical properties of the compound. Use of drug carriers.
3. Increase in cost-efficiency	Improvement of drug design and screening methods. Optimization of the synthesis pathway and technology.

[1] Here, we do not consider methods for prolongation such as drug encapsulation and use of plasters.

multidimensional space of various molecular characteristics is considered, within which the activity distribution of the compounds is analyzed. The key properties that could lead to the prediction of potential new-type drugs remain uncertain. The general approach consisting more of the molecular properties is considered the better. However, some properties are correlated to each other, and including them into any model would result in its poor consistency or multicollinearity.

The quality of collections of compounds (molecular libraries) acquired by pharmaceutical companies could be a critical factor in the search for new drugs. The quality of the compound collection is determined by the following characteristics [23]:

- number of compounds
- chemical variety determined by molecular skeletons and functional groups
- the extent of overlap of the collection of the compounds that can be acquired from other external sources
- number of compounds that are presumed not to react with nonspecific proteins

- number of compounds with structures unlike the structures of known pharmaceuticals
- number of molecular skeletons that are unlikely candidates to support the growth of drug-resistant microorganisms
- number of molecular skeletons or groups that do not cause undesirable side effects
- degree of purity of the compounds, polarity of the compounds, and stability of the compounds
- cost of the compounds
- possibility of testing and development of the drug-prototype.

Furthermore, collection should include drug-like (or bio-like) compounds. Numerous computer programs exist for compound selection. More than 140 million organic compounds are included in the chemical abstracts catalogue, from which 20–30 million compounds can be considered as potential drug candidates. In the 1990s, the pharmacological company Pfizer Inc. created a large collection of such compounds that were selected according to the developed criteria for drug-like compounds. The next selection of compounds for screening was performed. The probability of successful selection with the help of theoretical parameters was not higher than that in the case of random screening [24].

Chemical space of all possible small organic molecules is far larger and exceeds 10^{60} compounds [25]. Considering this vast amount, methods of construction and optimization of virtual libraries have become increasingly demanding, to put accent from improving selection requirements to checking the quality of the constructed virtual library. Therefore, it was proposed long ago to put much efforts on methods for making virtual libraries of de novo chemical compounds [26]. In the recent 5 years, with increasing efficiency of neural network algorithms, machine learning approaches to better explore the chemical space and construct chemical libraries with de novo compounds have been intensely applied [27, 28, 29, 30].

One of the selection algorithms assumes that the molecule skeleton (connected graph comprising the maximal path connecting all heavy atoms in the molecule) is the vital component responsible for its biological activity [24]. Based on the analysis of natural biologically active compounds, the selection focuses on cyclic structures. Next, the search for "missed" molecular skeletons is conducted. For this purpose, the skeletons are designed based on the prescribed cycle descriptors, and all possible virtual structures are generated by introducing heteroatoms and functional groups into the skeleton.

The quality of the constructed molecular library often becomes better with the decrease in its size. However, strategies for improving library quality (e.g., based on postprocessing docking) may be proposed [31]. Various criteria are used to screen virtual structures—from molecular mass (molecules with mass above 850 Da have often been considered bad drug candidates), synthesizability, and octanol–water partition coefficient, to more sophisticated selection criteria. Preference should be given to the groups most often found in the composition of effective drugs. The best 100 drugs in the current pharmacological market contain no more than 15 types of functional groups [13, 32]. It is worth mentioning that the possibility of missing highly effective compounds during the selection cannot be ruled out for all existing selection methods for competitive drug candidates. This possibility can be increased many times by adding considerations of cost-efficiency of the synthesis of one or more virtual compounds.

In rare occasions, the search for fundamentally new drugs can be targeted. Radioprotectors, which became a necessity because of the growth of nuclear power generation, did not exist 70 years ago. Investigation of the physicochemical mechanisms of ionizing radiation effects—generation of free radicals, peroxides, and other reactive chemicals in living organisms—resulted in the development drugs that inhibit the formation processes of these compounds. Antioxidants were not listed among the drugs several decades ago. These compounds have emerged because of the elucidation of the mechanism of cellular membrane oxidation. Antioxidants are inhibitors of the oxidation chain reactions that accompany many pathological processes in an organism [33, 34].

However, targeted drug design is rather an exception to the rule. Unfortunately, nowadays one cannot identify any strategy for the discovery of a new type of drug that demonstrates a significant predictive power.

12.3   SEARCH FOR DRUGS WITH IMPROVED PROPERTIES

The development of drugs with improved properties is a challenging task, but it is much simpler than the one described in the previous section. There is a certain algorithm to modernize the known therapeutic agent—that is, modification of the base structure. Numerous strategies have been invented thus far to improve pharmacodynamics and ADMET (absorption, distribution, metabolism, excretion, and toxicity) properties. A review of these strategies is out of the scope of this book, but the current state of *in silico* pharmacokinetics tools and methods enables the reliable prediction of ADME properties, which greatly diminishes the percentage of drug candidates removed from future development because of poor pharmacokinetic properties [35]. However, preclinical toxicity and clinical safety of the drug candidates are still major issues in drug design and development.

Very often, the increase in the compound's affinity to the target receptor achieved through the introduction of anchor nonpolar or slightly polar groups or fragments is accompanied by the increase in selectivity. However, sometimes, the opposite can also occur. In such cases, one can apply more complicated approaches such as by removing the elements of pharmacophores of nontarget receptors from the molecular structure or by decreasing or increasing the size and conformational mobility of the substituents to obtain a larger difference in the affinity to the target receptor when compared with the affinity to other acceptors.

The side effects of drugs include undesirable metabolic burden for the organism. In these cases, the increase in the compound's hydrophilicity and its excretion through the kidney in its intact form can be beneficial.

The adverse side effects should be classified into the ones associated with the interaction of the drug with nontarget receptors and the ones associated with intoxication due to blockade of the receptors responsible for the main pharmacological effect. For example, atropine overdose can lead to a psychopathological state because of the excessive blockade of the central mAChR. Therapeutically relevant doses mainly affect the peripheral mAChR, the blockade of which is associated with the antiulcer effect of atropine. One barrier preventing the elimination of the side effects of the drug is our limited knowledge on the pharmacophore structures.

The problem of increased tolerance to drugs is one of the most difficult problems in pharmacology. The mechanisms of increased tolerance are not limited to the interaction of a drug with its receptor, but they also involve different biochemical systems of the organism. Diversification of the therapeutic agents is not sufficient to fight the problem of increased tolerance. One of the causes of increased tolerance is associated with immunology. Both the therapeutic agent itself and its metabolite can form a conjugate with a protein, forming a hapten. Formation of the corresponding antibodies initiates an immune reaction in the organism, although weak.

Prolongation of drug action may be a less complicated task when compared with the problems discussed above. Prolongation can be accomplished by increasing the resistance of the compound to biodegradation masking unstable bonds in the molecule. Increase in the affinity, accompanied by the increase in the lifetime of the complex of the drug molecule with its receptor, can also lead to prolongation of the drug action. In several cases, the result can be achieved not by developing a new drug preparation, but by employing directed synthesis of prodrugs—compounds that are converted into the required active drug in an organism. Examples of prodrugs synthesized accidently are

- prontosil (transformed into sulfanilamide)
- phenacetin (transformed into paracetamol)
- codeine (transformed into morphine)

Currently, prodrugs are synthesized intentionally with the objective of prolonging the drug action and increasing the activity and selectivity of the preparation. Currently available prodrugs account for nearly 10% of all commercially available medications [36]. Of all the medications (belonging to low-molecular-weight compounds) approved for clinical application in 2008, prodrugs accounted for 33% [37].

It is worthwhile to mention the idea to overcome the low selectivity of antitumor drugs through its targeted delivery to the affected tissue [38, 39]. Unlike normal cells, cancer cells produce α-fetoprotein and contain its specific receptors. α-Fetoprotein conjugated with antibiotics is successfully used for the targeted delivery of the drug to the tumor. Effective inhibition of the experimental tumor growth was accompanied by the significant increase in the animal survival rate when compared with the control. Moreover, free antibiotic was significantly less effective than the conjugate.

If the effects of the drugs were associated only with their action on a single receptor, then it would be sufficient to have only a few drugs in each pharmacological class. However, dozens of drugs are used in each class. These compounds, as a rule, display their own set of features ("colors") as pharmacological agents. Coupling of neural subsystems with different mediator specificity creates opportunities to display "shades of colors." Drugs can affect different presynaptic processes and interaction with secondary messengers, change the number of specific receptors. The efficacy of a drug depends on the physiological peculiarities of an individual organism; for example, hormonal status. Personalized use of drugs is not too distant future.

It is not so rare occasion when the drug should or, vice versa, should not display a certain type of pharmacological activity based on its parameters; however, paradoxically, these effects are observed. One of the reasons for such behavior could be the competitive relationships of the compound with mediators, which manifested differently in different sections of the nervous system because of the different concentrations of the mediator in the synaptic cleft.

It is worth concluding with the following observation: there are three stages to estimate a drug's ability starting from drug innovation—first, the drug is considered as a panacea; second, it is treated as toxic; and third, it becomes one of the standard pharmaceuticals [6].

REFERENCES

[1] F. Ooms, *Curr. Med. Chem.*, vol. 7, p. 141, 2000.

[2] J. DiMasi, H. Grabowski and R. Hansen, *J. Health Econ.*, vol. 47, p. 20, 2016.

[3] O. Wouters, M. McKee and J. Luyten, *J. Am. Med. Assoc.*, vol. 323, p. 844, 2020.

[4] R. Rydzewski, Real World Drug Discovery: A Chemist's Guide to Biotech and Pharmaceutical Research, The Netherlands: Elsevier, 2008.

[5] M. Dickson and J. Gagnon, *Nature Rev. Drug Discov.*, vol. 3, p. 417, 2004.

[6] D. Laurence and P. Bennett, Clinical Pharmacology, Edinburgh: Churchill Livingstone, 1987.

[7] J. Hughes, S. Sees, S. Kalindjian and K. Philpott, *Br. J. Pharmacol.*, vol. 162, p. 1239, 2011.

[8] T. Riggs, *Lancet*, vol. 363, p. 184, 2004.

[9] M. Rawlings, *Nature Rev. Drug Discovery*, vol. 3, p. 360, 2004.

[10] J. Ng and L. Ilag, *Drug Discovery Today*, vol. 9, p. 59, 2004.

[11] A. Abbot, *Nature*, vol. 455, p. 1164, 2008.

[12] M. Diwan, A. Hamza, J. Liu and C. Zhan, *J. Chem. Inf. Model.*, vol. 48, p. 1760, 2008.

[13] D. Nicolotti, T. Miscioscia, A. Carotti and F. Leonetti, *J. Chem. Inf. Model.*, vol. 48, p. 1211, 2008.

[14] R. Carr, M. Congreve, C. Murray and D. Rees, *Drug Discov. Today*, vol. 10, p. 987, 2005.

[15] P. Hajduk and J. Greer, *Nature Rev. Drug Discov.*, vol. 6, p. 211, 2007.

[16] H.-J. Böhm, A. Flohr and M. Stahl, *Drug Discov. Today*, vol. 1, p. 217, 2004.

[17] Y. Hu , D. Stumpfe and J. Bajorath, *J. Med. Chem.*, vol. 60, p. 1238, 2017.

[18] H. Sun, G. Tawa and A. Wallqvist, *Drug Discov. Today*, vol. 17, p. 310, 2012.

[19] S. Melkko, C. Dumelin, J. Scheuermann and D. Neri, *Drug Discov. Today*, vol. 12, p. 465, 2007.

[20] R. Lerner and S. Brenner, *Angew. Chem. Int. Ed.*, vol. 56, p. 1164, 2017.

[21] M. Clark, *Curr. Opin. Chem. Biol.*, vol. 14, p. 396, 2010.

[22] B. Stockwell, The Quest for the Cure: The Science and Stories Behind the Next Generation of Medicines, New York: Columbia University Press, 2011.

[23] R. Wife and J. Tijhuis, In: *Drug Discovery and Development*, vol. 2, M. Chorghade, Ed., Hoboken: Wiley Interscience, 2007, p. 265.

[24] A. De Laet, J. Hehenkamp and R. Wife, *J. Heterocyclic Chem.*, vol. 37, p. 669, 2000.

[25] P. Kirkpatrick and C. Ellis, *Nature*, vol. 432, p. 823, 2004.

[26] G. Schneider and U. Fechner, *Nature Rev. Drug Discov.*, vol. 4, p. 649, 2005.

[27] X. Yang, J. Zhang, K. Yoshizoe, K. Terayama and K. Tsuda, *Sci. Tech. Adv. Mat.*, vol. 18, p. 972, 2017.

[28] Y. Yang, K. Yao, M. Repasky, K. Leswing, R. Abel, B. Shoichet and S. Jerome, *J. Chem. Theory Comput.*, vol. 17, p. 7106, 2021.

[29] M. Popova, O. Isayev and A. Tropsha, *Science Adv.,* vol. 4, p. eaap7885, 2018.

[30] W. Jin, R. Barzilay and T. Jaakkola, *arXiv*, vol. 1802, 04364, 2019.

[31] J. Lyu, J. Irwin and B. Shoichet, *Nat. Chem. Biol.*, Vols. https://doi.org/10.1038/s41589-022-01234-w, 2023.

[32] F. Burden and D. Winkler, *J. Chem. Inf. Comp. Sci.*, vol. 39, p. 236, 1999.

[33] E. Burlakova, A. Alesenko, E. Molochkina, N. Palmina and N. Khrapova, Biological Antioxidants in Radiation Damage and Malignant Growth, Moscow: Nauka, 1975.

[34] E. Burlakova, In: *Chemical and Biological Kinetics: New Horizons*, vol. 2, E. Burlakova and S. Varfolomeev, Eds., USA: CRC Press, 2005.

[35] M. Waring, J. Arrowsmith, A. Leach, et al., *Nat. Rev. Drug Discov.,* vol. 14, p. 475, 2015.

[36] K. Huttunen, H. Raunio and J. Rautio, *Pharm. Rev.*, vol. 63, p. 750, 2011.

[37] V. Stella, *J. Pharm. Sci.*, vol. 99, p. 4755, 2010.

[38] G. Posypanova, V. Makarov, M. Savvateeva, A. Bereznikova and E. Severin, *J. Drug Target*, vol. 21, p. 458, 2013.

[39] N. Yabbarov, G. Posypanova, E. Vorontsov, O. Popova and E. Severin, *Biochemistry (Moscow)*, vol. 78, p. 884, 2013.

Index

Note: Page locators in **bold** and *italics* represents tables and figures, respectively.

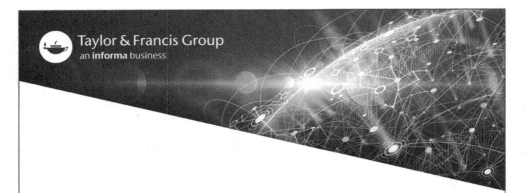